Translating Technology in Africa
Volume 1: Metrics

Translating Technology in Africa
Volume 1: Metrics

Edited by

Richard Rottenburg, Faeeza Ballim and Bronwyn Kotzen

BRILL

LEIDEN | BOSTON

The Library of Congress Cataloging-in-Publication Data is available online at https://catalog.loc.gov

Typeface for the Latin, Greek, and Cyrillic scripts: "Brill". See and download: brill.com/brill-typeface.

ISBN 978-90-04-67834-7 (paperback)
ISBN 978-90-04-67835-4 (e-book)
DOI 10.1163/9789004678354

This book is printed on acid-free paper and produced in a sustainable manner.

Contents

Acknowledgements

The series "Translating Technology in Africa" is largely based on the collaborative research programme "Adaptation and Creativity in Africa" (SPP 1448) funded by the German Research Foundation (DFG). The series was developed at the Wits Institute for Social and Economic Research (WiSER) at the University of the Witwatersrand, Johannesburg, South Africa. It is one of the outputs of WiSER's "STS Africa" research focus.

Figures and Tables

Figures

Tables

Notes on Contributors

Faeeza Ballim

is an historian and senior lecturer based at the Department of History at the University of Johannesburg. She holds an MSc in African Studies from the University of Oxford and a PhD in History from the University of the Witwatersrand. Her book entitled *Apartheid's Leviathan: Electricity and the Power of Technological Ambivalence*, was published by Ohio University Press in April 2023. Her main scholarly interest is to bring insights from Science and Technology Studies to bear on key themes in African history, initially through the consideration of dynamics in large energy-related infrastructure projects.

Kevin P. Donovan

is a Lecturer in African Studies and International Development at the Centre of African Studies at the University of Edinburgh. His book, *Money, Value & the State: Economic Sovereignty & Citizenship in East Africa*, will be published by Cambridge University Press in 2024. He has also written on economies of data and debt, infrastructure and surveillance, and corporate-state relations.

Véra Ehrenstein

is a full-time CNRS researcher in Science and Technology Studies based at the Centre d'étude des mouvements sociaux, Ecole des hautes études en sciences sociales (EHESS) in Paris. Her research explores the sciences and politics of climate change in a variety of contexts in Europe, Africa, and North America. She has written on topics such as carbon markets and tropical forest governance. She is the co-author of *Can Markets Solve Problems? An Empirical Inquiry into Neoliberalism in Action*, published by Goldsmiths Press in 2019.

Jonathan Klaaren

is Professor of Law and Society at the University of the Witwatersrand in the Law School and at the Wits Institute for Social and Economic Research (WiSER) in the Faculty of Humanities. He works in the areas of competition and regulation, the legal profession, migration and citizenship, and sociolegal studies. His most recent sole-authored book is *From Prohibited Immigrants to Citizens: The Origins of Citizenship and Nationality in South Africa,* UCT Press, 2017. He holds a Phd in sociology from Yale University and law degrees from Columbia University (JD) and Wits (LLB). In 2016, he served as an Acting Judge for the High Court of South Africa.

Bronwyn Kotzen

is a visiting research fellow based at the Wits Institute for Social and Economic Research (WiSER) at the University of Witwatersrand and a practicing architect. She holds a Masters of Architecture from the University of the Witwatersrand and an MSc in City Design and Social Science from the London School of Economics and Political Science. Bronwyn is a PhD candidate at the University of Cape Town and has lectured widely at multiple universities. Her research examines the sociomaterial politics of cement in Sub-Saharan Africa as a lens to explore her broader scholarly interests in the relationship between materiality, science, technology, and space. She has worked internationally in multidisciplinary teams on a wide range of award-winning projects across architecture, design, and academic research.

Emma Park

is an Assistant Professor of History at the New School for Social Research and Eugene Lang College, New York, where she teaches courses on modern Africa, science and technology, global histories of capitalism, and the history of development. Her current research focuses on infrastructure development projects including roads, radio and Kenya's now-famed telephonic banking service, M-PESA to explore transformations in capitalism and state-craft. Her book manuscript tentatively entitled, *Infrastructural Attachments: Technologies, Mobility, and the Tensions of Home in Colonial and Postcolonial Kenya,* challenges the seemingly apolitical nature of infrastructures along which state authority extends, binding citizens both to one another and to their state.

Helen Robertson

is a lecturer based at the School of Computer Science and Applied Mathematics at the University of the Witwatersrand, were she currently teaches logic, elementary theory of computation, and the applied ethics of data science. She trained in philosophy at the same university in Johannesburg and received her PhD from University College London. She has research interests in contemporary epistemology and the epistemology and ethics of artificial intelligence. She has guest edited a special edition for the *South African Journal of Philosophy* that focuses on the philosophical and ethical questions raised by the technologies of the Fourth Industrial Revolution.

Richard Rottenburg

is Professor of Science and Technology Studies at the Wits Institute of Social and Economic Research (WiSER), University of the Witwatersrand. His work has been inspired by renditions of post-foundational social theory and material-semiotic praxiography. The main objects of his inquiries are assemblages of evidence-making (experiments, quantifications), their dependency on knowledge infrastructures and their entanglement with narrative forms of sense-making and technopolitics. The driving questions are: How do these assemblages shape future-making and the attribution of responsibilities? How do they circulate around the globe? Rottenburg is best known for his 2009 book *Far-fetched Facts* with MIT Press.

René Umlauf
is based at the Centre for International Health Protection at the Robert Koch Institute in Berlin. He worked as a post-doctoral researcher at the Department of Anthropology at Martin-Luther-University, Halle, after receiving his PhD in Sociology from the University of Bayreuth in 2015. In 2020 he joined the Department of Sociology at Leipzig University where he started a project on humanitarian drone infrastructures entitled *Processes of Spatialization under the Global Condition*, at the Collaborative Research Centre. His research focuses on conceptual and methodological implications of technological change.

Helen Verran
holds positions as Professor at Charles Darwin University and an honorary position in History and Philosophy of Science at the University of Melbourne, where she taught for nearly twenty-five years. She has written many articles on numbers that ask about the work they do in different cultures and societies. Her book *Science and an African Logic*, University of Chicago Press, 2001, articulates the logic that Yoruba numbers express, showing it as profoundly different than that of modern numbers, which are derived from the blending of numbers originating in Indian, European, and Semitic cultures.

Translating Technology in Africa

Richard Rottenburg

1   Probing a Problematisation

It is generally understood that problems are not found, but need to be constructed. The first question is thus: how and what does this series on *Translating Technology in Africa* problematise? Together, the volumes in the series offer exercises in problematising the role of science and technology in Africa. In order to do so, the series probes the multiple ways in which science and technology are circulated, translated, and entangled with local practises. The relevant circulations are largely bi-directional and take place between temporalities (past to present), geographical spaces (urban to rural, country to country), and societal spheres. The entanglement of the technoscientific sphere with political, economic, legal, social, and cultural threads sits at the heart of each of the volumes and their chapters, evidencing the resultant particular and ever-changing assemblages (Suchman 2007). The collected case studies each describe concrete empirical events and situations that have manifested across various sites on the African continent. Most are variations and combinations of a few types of practices such as technoscientific solutions of practical problems, experimental innovations, melioristic interventions, creative adaptations, improvisations, and tweaks. Central to each of these practices lies the act of *translation*.

This problematisation immediately elicits a sceptical question. The practices listed are implicated in the overarching role of science and technology in preparing and supporting European imperialism and colonialism, and in the destruction of the planet's ecology. They are also implicated in the more hidden ways in which postcolonial forms perpetuate and increase intolerable inequities. This book series proposes an effective way of addressing exactly these disturbing aspects by first identifying a contextualisation and an entry point that circumvents binary juxtapositions between "us" against "them," "local" against "imported," or "familiar" versus "alien." Importantly, the series argues against essentialising and ontologising the characteristics of the juxtaposed entities.

This attempt at evading essentialising binaries follows a tradition that can be traced back to Leopold Sédar Senghor. During the first few decades after the independence of most African countries in the 1960s, the idea that "Europe underdeveloped Africa"—a phrase coined by Walter Rodney in 1972—implied a quest for development on the continent. This striving seemed indisputable in those years and was primarily directed towards technoscientific development. It was programmatically expressed by Senghor in January 1974 at the "Conference of the Ministers of African Member States Responsible for the Application of Science and Technology to Development," organised by UNESCO in Dakar (UNESCO 1974). The core political aim of the time was to compensate for what colonial governments failed to do and partly suppressed in their mission to establish technological infrastructures necessary for the extraction of mineral resources and agricultural products. The stance influenced by Senghor was later reinforced by Souleymane Bachir Diagne (2013) and Achille Mbembe (2021). According to the latter, the colony created zones of creolisation that foster exchange, dialogue, and new imaginations, and possibilities for an alternative version of global modernity for the planet. In Mbembe's rephrasing of an argument by Frantz Fanon it is the establishment of self-ownership through disalienation that will make "the creation of a new species of men and of new forms of life" possible (Mbembe 2021, 55). In this sense, our series of volumes on translating technology is interested in emergent shifts in the capacities of emancipation as related to the technosciences in postcolonial African contexts.

Such an approach prioritises two contextualisations: *circulation* (as opposed to inheritance) and *future-orientation* (as opposed to past-orientation). In other words, the entry point is not *origin* but *future-making*. The question is then how technoscientific knowledge production should be transformed to better serve a sustainable future for the planet—as the most precious common good—instead of serving capital reproduction. In short, how can technoscientific knowledge production be transformed to serve the ongoing search for the *commons* (Stengers 2015)?

Focused mainly on the highly technicised and industrialised countries of Europe and North America, Science and Technology Studies (STS) emerged in Europe during the 1970s as a distinct academic field with its own agenda. Over twenty years after STS was recognised, gained prominence, and became institutionalised in some of these countries—albeit on a very small scale—its problematisations and approaches were picked up in Latin America, Southeast Asia, and Africa where they were critically translated into postcolonial contexts and in so doing, modified. However, the level of institutionalisation and popularity of the field is lowest at African universities and research institutions. Broadly speaking, on the one side of the general academic campus is a robust quest for the modernist advancement of science, medicine, engineering, mathematics, and computation which is officially celebrated and well-financed. This happens largely unmoved by the scepticism towards the colonial imbrications of technoscience raised on the other side of the campus. Here, an equally robust quest for postcolonial critique of modernist and universalist understandings of technoscience has flourished in the humanities and social sciences, while reciprocally remaining largely unresponsive to the deep convictions of the scientists and engineers.

STS is an attempt to disturb this conspicuous configuration that can be found around the world by unearthing its inconsistencies and examining the entanglement of the technoscientific sphere with the political, economic, legal, social, and cultural threads. It does not reduce technoscience to power and money, nor power and money to technoscience. STS also rejects the assumption that technoscience is an expression of deep cultural assumptions or hidden laws of history to be revealed by cultural studies and philosophy. Rather, its objects of study are situated somewhere in the middle, in the details of the practices of the interstitial spaces between the two sides of the academic campus.

Another STS attempt to disturb an influential and seemingly stable understanding of science refers to the erroneous conviction that what matters is "high theory" and "breakthrough experiments" conducted by few chosen men of genius. Yet what really matters are the innumerable and often minute practices, their material infrastructures, and the complex ways in which they interweave to form temporary configurations (Elias 1992), with reference to some specific *zeitgeist*. Indisputably, there are theoretical and conceptual considerations aimed at making predictions that can be corrected (Chalmers 1976), mainly through experiments (including research instruments), measurements and statistical calculations (including digital technology). However, framing narratives such as progress and apocalypse, or equality and justice, and unpredictable intricacies of concrete situational practices also play an important role in shaping the production of technoscience. The latter often remain largely invisible.

The problem with this invisibility is that it significantly distorts public understanding of what scientific knowledge means. Invisibilising the contingency of the research process and the probabilistic nature of any scientific finding within the political discourse invokes a false notion of certainty. This invocation of certainty ultimately feeds either a dangerously naïve belief in science or an equally hazardously disbelief. The fact that things notoriously remain uncertain and often turn out rather differently than predicted does not justify any of these convictions. It rather proves that science is open to correction.

The authors of this series aim to critically explore technoscientific practises by holding these contextualisations in mind, as mentioned, and focusing on *future-making* instead of *origins*. Offering detailed descriptions of technologies at work, they engage with STS-inspired concepts and methodologies. They do so without assuming STS as a canonised body of superior scholarship imported from elsewhere, waiting to be deployed to the African context. They also do not assume that there is another predestined or grander way of knowing how technoscience and politics are related to each other. To the contrary, the authors aspire to critically and open-endedly unpack the existing scholarship and develop it further, so as to improve both Africa-related Social Studies of Science and Technology and STS in general.

This endeavour is important since contemporary governance around the world, including in African countries, often centres on regulating various technoscientific measures. The current COVID-19 pandemic underscores a configuration of social, economic, and environmental problems that were in fact glaringly obvious long before the pandemic. Most critical social problems of government—like equality, justice, health, education, and employment—relate to inequalities that are systematically engendered by capitalism. Simultaneously, these universal sociopolitical and juridical problems of capitalism relate to issues of the environment such as climate change, energy resources, and toxic residues. Both types of problems and their entanglement—and this is what the series is interested in—are framed and addressed by evidence that is produced within the technoscientific field. A comprehensive understanding of how this field works is therefore crucial for democratic participation, government, and corporate decision-making processes. Without a sufficient level of lay expertise, democracy is in danger of falling prey to flawed understandings of the technosciences and how they shape the world and are shaped by it.

It is equally important that technoscientific experts have an adequate understanding of the presuppositions within which their practices are enfolded, of the assemblages that their practices and material objects are unavoidably part of, and last but not least, of the popular understandings of what it is that they actually do. Considering this background, the series aims to trigger a vivid curiosity towards this field of inquiry. It intends to demonstrate that the different settings in which technoscientific solutions are developed and deployed require different approaches to examine them. It thus does not import STS to Africa but offer inspiration on how to rewrite STS in and from Africa—not only for students of the humanities, social sciences, and law, but also for students of the sciences, engineering, and medicine.

All these goals cannot be achieved without the active involvement of technoscience. Modernist technoscience emerged in Europe during the seventeenth and eighteenth centuries by translating diverse knowledges first from Europe and then from most parts of the world into a new assemblage. During this translation process a lot was lost and sacrificed for the sake of coherence. During the subsequent two centuries, this assemblage expanded around the whole world under the banner of progress and civilisation—a banner that regularly served as poor camouflage for exploitation and crimes against humanity.

Modernist technoscience did not prevent but rather amplified the mechanisms of self-devouring growth and ruination by the reckless industrial production not only of growing inequalities, but also of toxic residues that now threaten to terminate human life on the planet. Twenty-first century technosciences need to create new forms of sustainable future-making that will have to be measured against their ability to contribute to healing the planet. Achieving this goal presupposes that locally situated methodologies and technoscientific archives need to be recognised, represented, and included into the making of technosciences that can guide humanity out of the catastrophic times we are living through.

2   Probing Concepts

Against this backdrop, each volume in the series inquires into the multiple ways in which science and technology are translated so that they circulate and entwine with various other threads of concrete practices that result in ever-changing assemblages. Although the case studies examine different empirical events and situations, they all pay attention to the emergence and unfolding of ever-changing yet patterned assemblages, while simultaneously critically probing the analytical concepts they deploy.

The concept *assemblage* as used in much of STS literature is translated from the French word *agencement* which has no exact English counterpart. Agencement—the word agency is recognisable within it—refers to an arrangement or a combination of heterogeneous elements adjusted to one another, without the understanding that a human agent has arranged a number of passive things according to a human plan to expand human agency. Quite the contrary, agencements are endowed with their own capacity of acting. In the original understanding by Deleuze and Guattari (1987), there is nothing outside of an agencement because its description and the construction of its meaning is also part of it. This is to say that a sociotechnical agencement includes the language that describes it. To argue this way, Deleuze and Guattari need to contradict their own proposition as they claim to critically look through the recursive loop while maintaining that this is impossible.

One can, however, reasonably maintain that human sensemaking is able to self-reflectively include this loop into its operation as a way to inspect any sociotechnical assemblage from the outside. Insisting on this capacity of self-reflexivity does not mean restating the modernist claim that humans are in control of the assemblages of which they are a part. It equally rejects the opposite understanding that humans cannot be held responsible for the assemblages they helped to create.

Assemblages, as we have emphasized above, are the result of combinations of heterogenous elements adapted to each other. Some of these elements are human and others non-human, some are material and others immaterial. One can first think of simple examples like a library, a hospital, or an airline. At the same time and on a different level of classification, these combinations relate to differentiated fields of praxis—social, political, economic, legal, and cultural—with their distinct logics, codes, valuations, and normativities. One can easily imagine how the latter shape libraries, hospitals, or airlines. To employ a different vocabulary, we can speak of an entanglement of all the enumerated threads to form *material-semiotic assemblages* with their own agency. This agency is not situated in any of the threads themselves, but rather lies right in-between them. So far, we have seen how *assemblages* and *practices* are co-constituted. The assertion has been that they are eternally unfolding and never result in permanent patterns.

The one main cause for this perpetual change is prefigured in the fact that assemblages exist only as far as they are enacted in practices. Practices are patterned and thus imply repetition—otherwise one would not distinguish them from random activities and behaviours. Repetitions necessarily engender differences. What is repeated is being changed. As practices are therefore necessarily changing, they are—in the context of modernity—related to a horizon of changes for the better (Arendt 1998). One main route for trying out better possibilities for the future is to observe other peoples' practices in search of improvements of one's own practices. It is here that the notion of *circulation* gains importance for the approach that the authors of this series adopt. It is also here that it becomes clear why a search for *origins* fails to understand how assemblages unfold.

Assemblages and institutions cannot easily travel in their entirety and as such. This is due to the fact that they exist only as webs of relations between heterogeneous elements that constitute each other in the space between them. It is only some of the elements that can travel quite well. Think again of the three exemplar assemblages: a library, a hospital, an airline—none of which can travel. But books, medication and spare parts can easily be circulated. Librarians, doctors, and pilots can easily travel. Also techniques of cataloguing, diagnostic apparatuses, and repair and maintenance protocols can circulate. In order to depart, travel, and arrive,—in other words, to circulate—these elements disconnect from their assemblages and must be *translated* into mutable material-semiotic forms that depart and travel, and then adapt again to new forms that connect to other assemblages.

In this sense the central operation in the constitution of an assemblage is *translation*. This argument thus seems to imply that there is an *immutable* kernel to the *mutable* material-semiotic form that justifies the logic of translation. To reify the point, it helps to remember one of oldest meanings of the word *translation* as referring to the fundamental question of how far it was possible to "translate" from earth to heaven without death (Hebrews 11:5, Bible, King James Version of 1611). Notwithstanding its religious indexicality and implicit ontology—we are here speaking of the soul—this means that after long travels and many translations even an immutable kernel of a phenomenon will most likely get lost in translation. In the best case it will not die but *transmorph* into another form of existence.

Circulation and translation result in versions of *isomorphism* (things become similar) and with time of *transmorphisms* (a new form emerges). Firstly, isomorphism is often achieved through normative regulation like for instance, when all countries around the world agree to use the Gregorian calendar as a metacode for air traffic or stock exchange regulation, while they continue to use other calendars as their own cultural code. Secondly, isomorphism is often achieved through emulation or adaptation to the way others are acting; it is a form of mimetic learning, and hence the concept is also known more specifically as mimetic isomorphism. This happens, for instance, when many libraries, hospitals, and airlines around the world imitate a particular library, hospital, or airline that they believe is outstandingly successful. In such cases, a blueprint circulates and becomes a model through imitation. Without imitation there is no model. Those who imitate a blueprint create the model. Thirdly, isomorphism is often achieved by coercion. But in fact, the three versions of isomorphism (normative, mimetic, coercive) are ideal types, since dimensions of all three are often implicit in any real case. Over time, what once appeared to be isomorphism may turn out to be a case of transmorphism as the circulating form has been translated up to a point that a new form emerged which can start new cycles of imitation.

The European colonisation of great parts of the world including the African continent did not start as project of coercive isomorphism. The intention was not to install any of the political, juridical, and economic structures that were shaping European countries during those times. On the contrary, colonial extraction worked better by avoiding isomorphism and often began with brute force, executed with superior weaponry. However, as the European colonial powers pursued their economic interests, they were dragged into projects that implied coercive and normative forms of isomorphism into the field of for example, education, health, and basic infrastructure. Related projects aiming at normative isomorphism were joined by some forms of mimetic isomorphism on the side of colonised people who tried to make the best of the situation particularly in view of their liberation. One can say that European colonialism ended when a number of structures in the colonies had become isomorphic enough—like education, mobility, communication, and the form of the nation state—to organise effective resistance and liberation movements that would drive the colonisers out.

For the purpose of this series focused on *translating* technology in Africa, the relation between the *circulation* of technology and *trans-* and *isomorphism* is the main and central focus. Within this problematique, the relevant entry point here is that the travel of technology never completely follows fixed political boundaries of any kind including those of colonial or postcolonial states. Instead it results in differential outcomes that are also shaped by the spatial range of the technologies. The spatial extents of the resulting assemblages vary and do not yield coherent and clearly demarcated political spaces like national territories, regions, or continents. Some infrastructures are designed to integrate a specific territory by way of enclosure, such as policing and border protection for example, with their related technologies of surveillance and control. Other infrastructures are designed to cut across national territories to integrate regions like power grids, telephone systems, river bed management system. And finally, some assemblages are designed to span the globe such as the Gregorian calendar, the base ten numerical system, the decimal standard metric, and the internet and its protocols. In other words, if circulation leads to some form and level of technoscientific isomorphism, this does not coincide with political territories. It also does not coincide with older and historically stabilised technoscientific configurations and their archives which again do not coincide with entities such as "cultures" or "people." The latter are themselves shaped by changing technoscientific configuration.

To reiterate, a technoscientific archive is a cultural memory that entails many more technoscientific devices and procedures than an individual or a collective at a certain time in history actually uses. However, individuals and collectives who grew up within the range of a certain technoscientific archive would often recognise elements from it when, through some coincidence, they reappear from the hinterlands of their memory. As a test, think for example if you spontaneously know how the electric telegraph works? The key point here is that technoscientific archives do not correspond to territories like countries, regions, or continents, nor do they correspond to the extension of the realms of cultures, religions, languages, literatures, or narratives. Assuming that there were two (north/south), five (continental), or 206 (countries) clearly bounded technoscientific archives means to get the argument of the archive as epistemic space wrong. They are not like separate containers placed adjacent to each other as the territories of sovereign national spaces are. Rather, they cut across them. Archives overlap in various ways and are held together by "family resemblances": each technoscientific archive—like a family member—shares some traits with several other archives. No archive has unique traits and no single trait characterises all archives (Wittgenstein 1999, §65–71).

The authors of this series identify the ways in which the circulation and translation of technoscientific things—like protocols for experiments, units and formulae of measurements and statistics, methods of approximation, procedures, instruments, data collection models, data analysis, validations, case studies, and narratives—engender isomorphisms and transmorphisms. The hope is to contribute to a body of knowledge interested in future-making by following the principal question: how can technoscientific knowledge production become part of the *commons* to serve a sustainable future on the planet instead of optimising capital reproduction?

3   Probing Themes

The volumes of this series examine the unfolding of sociotechnical assemblages in different societal fields, in the present as well as in the past. The first volumes foreground different technoscientific equipment as part of various practices. Examples of these more specific themes are metrics and digitalisation; lifeworlds routinised by technoscientific equipment; infrastructures, procedures, and their users; devices and their users; technologies and the making of spatial configurations. Future volumes might include studies of technoscientific equipment related to biodiversity and interspecies relations. Together, the volumes aim to support and add momentum to ongoing and emerging debates about the role of technoscience both on the African continent and in general.

While these debates are characterised by unresolved controversies and open questions, this series gathers related detailed empirical work and theoretical reflections. We do not want to pretend that we can resolve all controversies and answer all open questions. We certainly do not believe that we can create a metaphysical tabula rasa to invoke a new ontology, or unearth an old one. Rather, we keep with Otto Neurath's interpretation: "There is no tabula rasa. We are like sailors who have to rebuild their ship on the open sea, without ever being able to dismantle it in dry-dock and reconstruct it from the best components. Only metaphysics can disappear without trace" (Neurath 1983, 92).

Bibliography

Arendt, Hannah. (1958) 1998. *The Human Condition*. Chicago: University of Chicago Press.

Chalmers, Alan F. 1976. *What is this Thing called Science? An Assessment of the Nature and Status of Science and its Methods*. Queensland: University of Queensland Press.

Deleuze, Gilles and Félix Guattari. (1980) 1987. *A Thousand Plateaus: Capitalism and Schizophrenia*. Minneapolis: University of Minnesota Press.

Diagne, Souleymane Bachir. 2013. "On the Postcolonial and the Universal." *Rue Descartes* 78, no. 2: 7–18.

Elias, Norbert. 1995. "Figuration." In *Grundbegriffe der Soziologie*, edited by Bernhard Schäfers, 75–78. Opladen: Leske and Budrich.

Mbembe, Achille. 2021. *Out of the Dark Night: Essays on Decolonization*. New York: Columbia University Press.

Neurath, Otto. 1983. "Philosophical Papers, 1913–1946." In *Vienna Circle Collection,* edited by R. S. Cohen, Marie Neurath, and Carolyn R. Fawcett, no. 16. Dordrecht: D. Riedel Publishing Company.

Rodney, Walter. 1972. *How Europe Underdeveloped Africa*. London: Bogle-L'Ouverture Publications.

Stengers, Isabelle. (2009) 2015. *In Catastrophic Times: Resisting the Coming Barbarism*. Lüneburg: Open Humanities Press and Meson Press.

Suchman, Lucy Alice. 2007. "Reconfigurations." In *Human-machine Reconfigurations: Plans and Situated Actions*, 259–86. Cambridge: Cambridge University Press.

UNESCO. 1974. *Science and Technology in African Development*. Paris: The Unesco Press.

Wittgenstein, Ludwig. (1953) 1999. *Tractatus Logico-Philosophicus. Tagebücher 1914–1916. Philosophische Untersuchungen*. Frankfurt am Main: Suhrkamp.

CHAPTER 1

Introduction

Richard Rottenburg

1 Metrics in Africa as Object of Investigation

For decades, most debates about how to create and secure political, legal, economic, and social systems that benefit the greatest number of people on the African continent have been largely dominated by one burning issue. The logic of the problem is this: improvements in social and legal systems are almost always seen as dependent on improvements in technological infrastructures. These improvements, in turn, are almost always seen as dependent on the diffusion of technologies. Advanced technologies have usually been developed elsewhere and need to be translated into local contexts in order to be useful. In most of these translation processes, only some aspects can be adapted to local conditions, while others must remain unchanged for the technology to work. The technological limitations of adaptation lead to a degree of global standardisation and centralisation. In many cases, this is not conducive to the original purpose of translation. There is therefore an inherent tension between globalisation and localisation.

The technology at the heart of this volume is metrics. Nowadays, the term metrics usually refers to digital metrics, which in turn depends on digital infrastructures. These are characterised by the fact that they are global but do not have a single centre. They are deliberately run through decentralised networks. This also means that while the core elements of the infrastructure may be centralised in Silicon Valley, Shenzhen, or Hyderabad, they can be used in a decentralised way for local purposes anywhere in the world. At the same time, key elements of the definition of equivalence must be defined in the centres of computation in order to work globally. This strongly shapes and constrains the possibilities that remain in local contexts.

The contributors to this volume use a variety of empirical case studies to explore the tensions between centralisation and decentralisation in the translation of circulating forms of metrics. Examining mathematical operations among the Yoruba people of Nigeria, Helen Verran asks how different number systems and ways of calculating can coexist within the same community of practice. Helen Robertson examines the extent to which a machine learning model can be said to possess the relevant concept that precedes the classification which the machine is supposed to learn. René Umlauf tackles a related phenomenon by examining how machine learning models developed by a company in San Francisco are trained in a "data factory" in Uganda. Looking at digital lending in Kenya, Emma Park and Kevin P. Donovan ask how the centralisation and standardisation that accompany digitisation affect social variations in local lending practices. Véra Ehrenstein addresses a connected issue in the field of forest metrics in Gabon, asking to what extent centralised and standardised forms can become independent of local practices. Jonathan Klaaren examines how locally hard-won rights of access to information in post-apartheid South Africa have been shaped by the introduction of new information technologies, but continue to be influenced by their history.

The aim of this introduction is to provide a framework for a better understanding of the common themes that unite the very different topics of the chapters in this volume.

2   Situating Metrics

In order to further specify the focus of this volume and its objects of investigation, it is helpful to ask how metrics relates to non-numeric forms of knowledge practices that seek to make sense of the world by reducing its complexity.

Metrics is used as a generic term for various forms of numerical representation, also known as quantification, which includes counting, measuring, and calculating. Metrics seeks to grasp the general by analysing the relationship of the one to the many. It can be practised in analogue and digital forms, and in principle both follow the same mathematical logic (Didier 2021). In our time, metrics is most often associated with digitalisation, and so it is in this volume. While digitalisation does not change the logic of metrics, one of its effects, namely datafication and the resulting big data, seems to change it (Mayer-Schönberger and Cukier 2013). When interoperability between different data sets is achieved, the data collected can be used for different purposes that were not known at the time of collection. And once this state of affairs is established, the interest in installing technological infrastructures to generate data becomes an end in itself. Data mining and related new digital technologies are born and change the character of metrics.

Hermeneutics is used as a generic term encompassing all non-numerical forms of interpretation and representation, but above all it relates to language and words, to narrating and reading the world. Hermeneutics seeks to grasp the general by interpreting the relationship of the part to the whole, of the particular to the general. To some extent separately, but in principle together, metrics and hermeneutics constitute the world as we know it. Both kinds of knowledge practices generate their own apparatus of inquiry and archive of knowledge. Raising questions about the entanglements between metrics and hermeneutics is one of the main aims of this introduction, in order to frame the work of the following chapters.

The starting point for this volume is that metrics has become the most robust form of evidence in any public dispute around the world. This is due, at least in part, to the understanding that quantification translates the political into the technical and thus provides the most effective language for communication across social, cultural, and disciplinary divides. As early as 1904, Max Weber had a dark premonition of the modern rationalisation of the social world, and of quantification as an important dimension of it. He recognised that modernity was constructing a "steel-hard casing" for itself, and this filled him with a remarkable unease (Weber 1972, 203). Weber identified the elements used in the construction of this casing as Protestantism, capitalism, industrialism, bureaucracy, and mechanical technology, which together rationalised the Western lifeform. He feared that this very rationalisation would become a self-devouring process that would not end until "the last ton of fossil fuel has burned up" (Weber 1972, 203).

Since the beginning of the twentieth century, Weber's premonition has continued to haunt many public discourses. It has taken on an even darker tone with the event of digitalisation and datafication. In March 2023, the artificial intelligence community's call for a six-month moratorium on AI development to prevent chaos is probably one of the most extreme expressions of the dark premonition Weber had more than a hundred years earlier. The opening paragraph states:

> AI systems with human-competitive intelligence can pose profound risks to society and humanity, as shown by extensive research and acknowledged by top AI labs. As stated in the widely-endorsed Asilomar AI Principles, Advanced AI could represent a profound change in the history of life on Earth, and should be planned for and managed with commensurate care and resources. Unfortunately, this level of planning and management is not happening, even though recent months have seen AI labs locked in an out-of-control race to develop and deploy ever more powerful digital minds that no one—not even their creators—can understand, predict, or reliably control. (Future of Life Institute, 2023)

Of course, the problem is huge and the ability to see a pattern in its development is limited. As usual in such situations, not everyone is equally alarmed.

Some are more fascinated by the current panic and the escalation of a rhetoric of finality, arguing that these reactions are not primarily related to what is actually going on. Others are fascinated by the amount of attention, energy, and money being invested in post-solutionist projects, such as the search for an alternative planet as a future habitat for a humanity that has destroyed its original planet. Still others are convinced of the need to pay more attention to details and differences, and to avoid attributing truth to grand assumptions about the course of history when not enough is known. The contributors to this volume are mainly invested in the latter approach.

Closer to the questions raised in this volume, a growing number of concurrent studies problematise the implications and consequences of the expansion of digital metrics to countries with comparatively new and sparse networks of numerical representation, most of which are postcolonies (Breckenridge 2014; Jerven 2015; Nyabola 2018; Lamoureaux et al. 2021). Three fundamental questions come to the fore differently in sparsely-measured postcolonies than in densely-measured, highly-technicised countries. One question is about the extent to and the way in which trust in digitised metrics depends on trust in the organisations that produce the numbers and in the institutions that regulate the producers. We are dealing here with a particular version of the foundational circle of trust creation. As noted above, numerical evidence has become the golden standard of evidence in general. You need numbers to challenge numbers. But what can you produce as evidence if you do not trust the way the numbers are produced? A second question concerns the problem of data sovereignty in relation to state sovereignty. Finally, a third question problematises the tendency in African countries to move towards digital metrics without first institutionalising basic infrastructures of quantification. Taken together, these three questions seek to open up a space of inquiry that does not start from unproven assumptions about the adoption of digital metrics as a promising form of leapfrogging.

The contributors to this volume scale these larger questions down to a level at which they can be examined empirically through praxiographies. They follow Thomas Eriksen's credo of studying large issues in small places. In the remainder of this introduction, I argue for what the social study of quantification has to say about knowledge practices that depend on digital metrics in postcolonial African settings. The guiding question problematises the idea that doing metrics and asking about the deeper meaning of life are two separate endeavours. In this vein, I argue that rather than using hermeneutics to critically examine metrics, or using metrics to discard hermeneutical insights, it is perhaps more accurate and promising to ask how metrics and hermeneutics are intertwined (Didier 2021; Morgan 2022).

3   Hermeneutics

Hermeneutics maintains that the most elementary part of social life is the meaning related to acting—here understood in the emphatic sense of Max Weber's notion of *handeln* (Weber 1973). Actions are conceived as being meaningful to the participants of an interaction. Subsequent actions are understood to be oriented towards the meanings of prior actions conveyed through various forms of signification, mainly through narratives. Accordingly, to understand a particular event or situation one needs to look at the social actions that led to the particular event or situation and identify the meanings that the various participants give to their own actions and those of others acting before and after them. This implies the necessity to look at the institutionalised forms that are indispensable for sense-making and at the notions implied in these forms (Garfinkel 2012).

Some of the most basic concepts that are essential for making sense are causality, the principle of non-contradiction, temporality, rationality, responsibility, person, notions of good and evil, reciprocity, life, death and the afterlife, and all sorts of classifications and orders of things. But these concepts and the institutionalised forms associated with them—and this is the important point—cannot be used as sufficient explanations of action. Unlike behaviour, action by definition means acting freely and in view of making a difference. So, on the one hand, institutionalised forms are there to guide actions in certain ways and to prevent them from going otherwise. But on the other hand, because institutionalised forms are constantly re-enacted and only thus remain alive and functional, each repetition can—and often does—make a small difference. People mostly do not blindly follow rules and just behave, but they act to make a difference for themselves and others. Over time, forms therefore change and take on different meanings. In this view, action and form co-constitute each other in endless circles (Arendt 1998; Joas 1996; Boltanski 2011).

Rather than trying to escape this circularity, hermeneutics embraces it, claiming that the humanities must move in the same circles. In other words, the social sciences, as part of the humanities, must search for interpretations of the meanings of previous actions. This methodology is followed intuitively, for example, when reading a text. You take in the meaning of individual words, then the sentence of which they form part, and this in turn changes your original understanding of the meaning of the words. The same process is repeated for any larger semantic unit to which the words and phrases belong. Understanding a chapter, a book (or a film), a genre, an era, and so on, makes it possible to go back to a single word and realise a different meaning. These endless back-and-forth movements are known as hermeneutic circles. Classical hermeneutics, similar to classical phenomenology, conceived meaning as something purely ideational that can become attached to material objects but only due to human attribution.

This position resulted in the understanding that quantification can only deal with external things that have primary qualities like extension, shape, strength, number, degree of mobility, and due to these qualities, they can be counted, measured, and experimented with. In contrast, the ideas of the mind have none of these qualities and therefore do not immediately belong to the realm of quantification. In his seminal book *The Crisis of European Sciences and Transcendental Phenomenology,* Edmund Husserl (1970) critically diagnosed a loss of reality due to "Galileo's mathematization of nature" (Husserl 1970, 23–59).

This old fear of losing the true meaning of life and the essence of nature through the penetration of quantification into all areas of knowledge production is alive and well today. But so is the fascination with the possibilities that only numbers and mathematics allow. This ambivalence, the simultaneity of anxiety and fascination, has become more alarming by the increased importance of digital metrics in shaping contemporary lifeforms around the globe, including the increasing disparities between social spaces, countries, and continents.

The contributors to this book argue that the anxiety and fascination with the growing importance of digital metrics is not the result of clearly separated practices of knowledge, one of which is about to make the other redundant. Rather, they move closer to concrete instances of practice, looking at the details and examining if and how metrics and hermeneutics are intertwined. Two of the authors are more explicit about the difficulty of juxtaposing metrics and hermeneutics. In her chapter, Helen Robertson shows how numbers and meanings are intertwined in ways that raise fundamental doubts about the possibility of a strict separation of hermeneutics and metrics, let alone the possibility of replacing one with the other. Human medical experts around the world can, for instance, distinguish a benign from a malignant tumour. This is because they all possess the same relevant concept for diagnosing a tumour after the appropriate medical training. They hence understand the meaning of "malignant" and "benign" and can apply the words appropriately. Computers and the machine learning models that are run on them can also be used to distinguish benign from malignant tumours based on measurements and thus can assist the medical expert's decision-making process. However, Robertson shows that there is an important difference between the human medical expert and the computational model. The model does not understand the meaning of benign and malignant, but simply assigns certain measurements to the benign category and others to the malignant category. More specifically, Robertson shows that, even according to the most uncommitting account of what it is for a human to possess a concept, machines cannot be said to possess any concept s. The capability to possess a concept, that is, to understand meaning, is limited to human beings. It cannot be delegated to digital machines that only measure, count, compare, and calculate.

Helen Verran's chapter demonstrates that numbers, the basis of metrics, are already socially and culturally embedded. She examines the linguistic and arithmetic practices facilitated by two distinct numerical systems that emerged historically among the Yoruba people. One of which is associated with trans-Saharan trade, using cowrie shells as currency. The other is associated with British colonisation and schooling. Recognising that the two systems, with their different numerical forms, are based on different commitments, carry different social values and enact different relationships, the chapter asks how the two numerical systems relate to each other. In answering this question, the chapter refutes the common belief that numbers and mathematics—as they are known and taught in schools and universities around the world—stand outside any sociocultural fabric and carry reality itself in their forms.

4 Metrics

Over the last few decades, scholars of the history and philosophy of science and technology have examined the implications of numbers and calculative practices for the construction of knowledge. It is not surprising that interest in this topic arose in countries where quantification first emerged as a distinct and important area of practice, encompassing all social sectors notably in Great Britain, United States, France, and Germany. Research on the history of statistics highlights how this form of knowledge enables a new way of grasping and shaping the social world by being premised on a particular notion of objectivity (Porter 1986, 1995; Desrosières 1998). As Hacking (1975; 1990) shows, since the nineteenth century statistics and its probabilistic forecasting were fundamental to the rise of the modern nation state and its practices of governance in Europe and North America. Studies of the history of the notion of objectivity as it has emerged in modern public, political, and bureaucratic life have well revealed how numbers have come to signify an almost taken for granted understanding of impartial and objective knowledge (Daston and Galison 2007). To highlight this point, Theodore Porter has coined the phrase "mechanical objectivity" (Porter 1995).

This work is complemented by research on accounting and auditing that reveals the growing importance of numbers also in everyday life. In further developing Foucault's notion of governmentality (Foucault 2006; 2006a), scholars argue that accounting is not a purely technical and neutral practice as it fosters forms of disciplinary power (Miller, Hopper and Laughlin 1991; Miller 2001; Hopwood and Miller 1994; Espeland and Vannebo 2007). Social studies of accounting argued that during the 1980s countries at the forefront of neoliberal reforms were facing an "audit explosion" (Power 1997). The drive to compare public service delivery in order to thereby assess its relative cost-efficiency, led to stricter practices of auditing that in turn served to test the legitimacy of governing practices (Poovey 1998; Carruthers and Espeland 1991). Like the literature on statistics, the one on accounting raises concern that the increasing importance of quantitative evidence has created a situation where only those operations that are quantified are taken into account at all and that in this process, long-term goals have largely been lost from view. Narrative modes of knowing, so the literature argues, receive less attention even though they are deeply entwined with quantitative modes of knowing (Espeland and Sauder 2007; 2016; Morgan 2022).

Social studies of standardisation provide another valuable framework for considering the role of numeric representation in governance and everyday life. They show how technologies of quantification and formal representation (mathematical formulae and models, charts, graphic depictions) become indispensable in processes of modern rationalisation (Berg 1997; Bowker and Star 1999; for organisation theory, see also Brunsson and Jacobsson 2000; Morgan 2012).

The literature on statistics, probabilism, objectivity, accounting, and standardisation also shows that the production of numerical knowledge depends not only on its own relevant methodologies and research technologies. It also requires social, political, economic, and legal support, including networks of scientific peer support, financial resources, legal approval, and political recognition. Some of these supportive networks gradually become institutionalised and provide the basis for public trust in numeric representation (Shapin and Schaffer 1985; Porter 1986; Bloor 1991). Some social studies of quantification further emphasise that the meaning, purpose, and intended and unintended effects of institutionalised numerical representations differ across social domains, and therefore require careful empirical analysis of the proliferation of numbers and computational protocols within their social contexts (Callon, Millo and Muniesa 2007; MacKenzie, Muniesa and Siu 2007; Mugler 2018; Didier 2020; Didier 2021). In sum, at least since the 1990s, there has been a well-established and still growing body of scholarship that examines the expansion of metrics in all spheres of social life, while at the same time questioning the independence of metrics from hermeneutics and vice versa.

5   Doing Metrics

As elaborated in the previous section, it is generally accepted that modernity is inconceivable without numeric representation. Statistics and accounting have emerged as key forms of knowledge production and technologies of governance of industrialised states; probability theory, random sampling, market ideology, and the democratic welfare state have collectively co-evolved around the notion that independent agents choose freely and yet—in aggregate—predictably (Krüger, Daston, and Heidelberger 1987). From its beginnings, modernity created an affinity between governance and evidence accessible to the public, as Foucault's (1973) work authoritatively demonstrates by spelling out Friedrich Nietzsche's programmatic assertion.

> In order to have that degree of control over the future, man must first have learnt to distinguish between what happens by accident and what by design, to think causally, to view the future as the present and anticipate it, to grasp with certainty what is end and what it means, in all, to be able to calculate, compute—and before he can do this, man himself will really have to become reliable, regular, necessary, even in his own self-image, so that he, as someone making a promise [...], is answerable for his own future! (Nietzsche 2006, 36)

The necessary evidence implied in this capacity to become answerable for one's own future turned more numeric during the twentieth century. In contemporary democracies, most relevant questions—like where to build a road, a railway, a school, a hospital, a waste disposal site, a nuclear power station, or how and for what ends to use taxes, or who should be qualified to receive a credit—invite answers based on numeric evidence.

Through quantification, the world becomes knowable at a distance, neatly compartmentalised, and ordered. Things that at first appeared incommensurable can be made commensurable. While one always loses some aspects of the reality in question through numeric representation, one equally gains others that were invisible before quantification. The discovery of this can be attributed to the early French statisticians who at the beginning of the nineteenth century were greatly concerned with normality and deviance (Hacking 1990, 64–104).

Numeric representation lends itself to the generation of comparisons and rankings of known phenomena, but it also allows us to re-arrange data originally collected for a particular purpose into endless new configurations that enable the detection of previously unanticipated interconnections. Established forms are mostly simple, unambiguous, and seemingly easy to prove and understand. Once quantifications are established, they successfully hide the theoretical and normative assumptions inscribed into them. Against much of their public image, all forms of quantification do not mirror reality but are instead the product of a series of interpretive decisions about what to quantify, how to categorise, and how to label things. The more diverse and less countable the phenomenon being quantified, the more difficult and cumbersome these decisions are. New quantifications always rely on previous ones and are thus shaped by their logic and the kinds of data they generated.

Quantification also needs substantial resources and these depend on what governments and private organisations consider worth knowing in numeric form. What ends up being quantified, and thus encoded in particular ways, is often the product of what is understood as being problematic by relevant and influential publics. The very act of numeric representation constrains the kinds of information that are available. While selected problems and their connections that previously were hidden are made visible through refined modes of quantification, others remain concealed or become even more invisibilised by the dominant numeric representations. However, the power of numbers has reached a point where attempts to question dominant quantifications are themselves often presented in numeric form (Merry 2006; 2011; Hetherington 2011; Bruno, Didier, and Prévieux 2014). For the argument of this introduction, it is important to outline those operations of quantification that are conventionally contrasted with hermeneutics, yet should rather be interpreted as interwoven with it—as I propose here.

Central to any system of quantification is commensuration and comparison (Espeland and Stevens 1998; Heintz 2010). In order to collect data, it is essential to make things commensurable: to decide on a principle of similarity so that things can be grouped or classified, counted, and calculated. Only then can data be transformed into information, and then information into knowledge. The starting point is often a list or series of items that are easy to count. A logical first step is to establish some equivalence between all the items on the list, including variations that are not on the list but could potentially exist. This requires finding a commonality—a shared characteristic—between the individual cases and ignoring the differences (Desrosières 1998, 10–11). Once this construction is accepted, as in the case of the idea of the "average man," the common feature becomes a real thing. In this sense, we are dealing with a fundamentally performative practice which as such has predictive power (Osborne and Rose 2003; Didier 2020). By establishing such equivalences, categorisations (also known as classifications or taxonomies) are created. These, in turn, are defined and organised into a system of multiple categorisations, so that all things that seem relevant fall into one category or another, ideally mutually exclusive and together all-encompassing.

The creation of categories for the purpose of statistics and governance is, after all, an arena of significant interpretive work, shaped by pre-existing categories, theoretical concerns, and practical purposes. Debates recur in the history of statistical classifications about "a sacrifice of inessential perceptions; the choice of pertinent variables; how to construct classes of equivalence; and last, the historicity of discontinuities" (Desrosières 1998, 239). Taxonomies bring together things that do not necessarily belong together and attach a common label to them so that they constitute a single category. Moreover, each category has to be usable in all future situations in which the taxonomy is meant to order things in a meaningful way and thus has to be even more abstract than a given context already requires.

For the categories to be useful, they must be populated by individual cases that again need to be encoded into them. The encoding process refers to the decision to attribute an individual case to a particular class. For the argument about the unavoidable interlacing of metrics and hermeneutics it is important to emphasize that the act of classification is hermeneutic work that needs to downplay certain aspects of the case and highlight others. Here cultural, normative, social, political, and technical dimensions play important roles that are quite independent of the object to be encoded (Mervis and Rosch 1981.) In the end it appears as if objectively given things were simply sorted out and quantified, when in fact these things only become real in a certain way as a result of having been encoded. Over time and with use, they become more established and accepted as unquestionably real. "When the actors can rely on objects thus constructed, and these objects resist the tests intended to destroy them, aggregates do exist—at least during the period and in the domain in which these practices and tests succeed" (Desrosières 1998, 101). In other words, it is their institutionalisation that makes metric aggregates real and thus trustworthy.

Perhaps the most important part of this stabilisation through institutionalisation is the increasingly dense, all-encompassing, and complex web of cross-references between numeric forms of world-making. While one particular form of quantification might try to be as comprehensive and differentiated as possible towards a particular issue—like for instance health, unemployment, suicide, poverty, air pollution, racial discrimination—it unavoidably cross-references several other issues and strongly depends on the availability of those metrics. This means that a particular instance of a metric representation, such as one depicting the health of a population, is heavily dependent on the availability of several other metric representations, such as vital statistics, income, education, infrastructure, environmental aspects, workplace studies, and so on. Taken together, these different metric representations are much more robust and useful than they would on their own. This is one of the key differences between countries with scattered measurement networks and countries with dense measurement networks.

Whether scattered or dense, a self-stabilising web of numeric representations is distinguished by its specific and intended shallowness. It is a web of "thin descriptions" (Porter 2012). As such it contrasts with the web of narrative representations of reality characterised by its intended depth. Those are webs of "thick descriptions" (Geertz 2017). However, as I try to show in this introduction, thick and thin descriptions are each enfolding the other, without ever becoming completely subsumed or purified from each other. In the next section I delve into more details of quantification to strengthen this insight.

6   Forms of Digital Metrics

Metrics, as shown in the previous section, is never neutral, but always a form of technopolitics. Neoliberal governance as it emerged in the 1980s—first in North America and the UK, now everywhere—introduced a specific form of metric technopolitics. The innovation was not about unleashing seemingly eternal and natural market mechanisms that had been restrained by the state. It was about introducing measurements that created new market mechanisms. One of the main forms of neoliberal measurement, known as "benchmarking," aims to improve performance by comparison through indicators in contexts where there are no conventional market mechanisms to perform the same function (Bruno and Didier 2013; Mennicken and Espeland 2019; Mennicken and Salais 2022; Guter-Sandu and Mennicken 2022).

Benchmarking through digital metrics is officially linked to the idioms of subsidiarity, self-monitoring, self-auditing, and self-responsibility. In a discourse of supposedly increased civil liberties, control became largely a matter of self-control and was interpreted as a shift away from old structures of domination that privileged the few towards more democracy, freedom of choice, participation, and transparency. But as it turned out a few decades later, the resulting new order dramatically increased inequalities within and between states around the globe (Piketty 2014). And, as some scholars argue, new technological and other developments seem to have begun to reshape or end the neoliberal era since the new millennium.

Even in the neoliberal era, not all metrics were driven by the logic of benchmarking. Other forms of metrics continued to operate and new ones began to emerge. To understand the mechanisms at work, it is again necessary to pay attention to concrete practices and to differences between fields and sites of practice around the world. One of the more important dimensions of neoliberal reforms is that a large proportion of metrics is no longer produced by state institutions but by private or at least independent agents. To some extent this is related to benchmarking and performance-based funding, as in the case of privatised railways, postal services, telecommunications providers, other utilities, hospitals, libraries, universities, and sometimes even prisons. On the other hand, a new and opposite trend is emerging as a result of the same neoliberal intervention. Contrary to the neoliberal programme that sought to turn citizens into self-entrepreneurial subjects, contemporary forms of digital metrics reinvent the possibilities of centralised action. The central actor does not necessarily have to be the state, it can be a private corporation, an international organisation, a foundation. Often it is a private firm that is subcontracted by a state institution that thereby becomes fully dependent on the development and maintenance of the software to do the digital metrics for the state.

For the argument of this introduction, the crucial point of currently emerging forms of digital metrics is the creation of an "experimenter" who can analyse a "population" to objectify its behaviour and make it more predictable—as has been the case with conventional statistics since its inception. In this seemingly familiar context, however, new forms of digital metrics have come to the fore. One of these is the design of sociopolitical interventions as controlled experiments. They start from a known and given state of affairs that is seen as problematic and move towards identifying previously unknown relationships. The HIV/AIDS pandemic became the prototype of this form. Initially, only the symptoms were known, but later statistical analyses revealed previously unknown correlations between the symptoms and social variables. This helped guide medical research into the causes of the symptoms and led to the identification of the virus. Once a potential treatment was identified in the laboratory, it was quickly applied on a large-scale before going through the previously established and accepted medical procedures to measure efficacy and safety. The chronology from trial to treatment was partly suspended so that the treatment remained part of the trial. This particular approach was accepted as the norm in many fields outside the HIV crisis (Rottenburg 2009). It was further institutionalised in attempts to contain the 2014 Ebola crisis in Sierra Leone, Liberia, and Guinea, and quietly established as the gold standard during the COVID-19 pandemic.

Another version of the intervention-as-trial is used when the problem, its causes and the desired end are known, but the means of achieving the end are not. In most cases in this category, the means are desired changes in the behaviour of relevant groups. The relevant population, or a sample of it, is divided into two statistically equivalent groups, and a particular incentive for behaviour change is tested. For example, one village receives a water treatment system and the other does not. Differences that emerge after the defined trial period are then attributed to the presence or absence of the incentive. The Abdul Latif Jameel Poverty Action Lab at MIT provides good examples of this approach.

A key difference between large-scale modernist government interventions designed to make the world a better place and these newer interventions, which are run as experiments, is the way in which they conceive of the future. Based on the narrative of progress, high modernity interventions assumed that the future was partly indeterminate and therefore malleable. They aimed to make the future not only better but also more predictable by shaping it. In contemporary interventions, faith in the ability to shape the future has been replaced by a heightened and nervous awareness of the risks involved. It is largely for this reason that some interventions are designed as experiments to generate the data needed to correct the next phase of intervention, which is itself designed as an experiment (Ezrahi 1990). And yet some of these experimental interventions raise fears that they might grow to be out of control. Voices calling for political regulation based on ethical principles are becoming louder and more influential. They question the legitimacy of interventions that may have unpredictable consequences and demand that only those interventions should be pursued whose consequences can be reversed in the future (Jonas 1979; Jasanoff 1994; 2020). Digital metrics, which is always part of the problem and part of the solution, is needed to assess the relevant probabilities for both positions.

One important form of digital metrics remains largely unaffected by the spread of experimentality, though. It is mostly employed when a problem arises that seems expansive, pervasive, and hard to delineate. The exact articulation of the problem, its expressions, its causes and effects, and ultimately its remedies are unknown. A quintessential example of this was the premonition of climate change, which held a controversial place in the debates of the 1980s, yet the scientific controversy was in fact closed during that decade mostly due to successful metrics. Since then climate change, like poverty alleviation, is among those issues where the causes, effects, and even the remedies are well known. Although most of these remedies are in fact non-controversial in terms of their causal relevance (like the reduction of energy gained from fossil fuels), they are still hard to translate into workable and globally enforceable interventions because they demand substantial changes of the dominant lifeform particularly in the industrialised countries. Working out these interventions requires more metrics.

Searching not for specific cause and effect relations or statistical regularities by following a hypothesis, but for radical changes of the contemporary dominant lifeform is an open exploration. It is mainly about discovering surprising and potentially useful patterns in a huge pile of big data, without anticipating what those patterns might be. Hope is invested in previously completely unknown escape routes. This is one of the practical situations in which conventional metrics morphs into digital metrics associated with big data, data mining, and machine learning. In this kind of open exploration, digital metrics seems less like a tool of governance and more like a version of basic science. It is often run by centralised, permanent infrastructures that are set up for a specific problem, but with an open exploration focus. They collect, aggregate, and correlate large, heterogeneous sets of digital data and develop new models with open, non-experimental research questions. The National Oceanic and Atmospheric Administration (NOAA) (2021) is one and the European Centre for Disease Prevention and Control (ECDC) (2021) is another, among many examples.

Open explorations are increasingly also run by a rather different type of emerging infrastructure that is decentralised, non-hierarchical, participative, and facilitated by the rapid development of the Internet, and the increasing speed and capacity of computing. Web2 platforms facilitate mass participation by ordinary citizens, according to the wiki-principle and generate a new type of information gathering and new chances for quantification. A successful example of this type is the "Extreme Citizen Science Blog" (2021) housed by the Department of Geography of the University College London (UCL). While these new forms of lay expertise in processes of digital metrics are partly an expression of an increasing scepticism towards the type of institutionalised numeric expertise implied in political decision-making, they simultaneously reinforce the power of digital metrics, and contribute to its spread across all sectors of life.

7   Metrics and (De)Centralisation

Digital metrics is often celebrated for its improved calculations of probability for all sorts of developments—from the unfolding of a pandemic, over
environmental and climate changes, to economic and demographic transformations and any kind of social phenomena, including positive effects of digital technologies and robotics for society. Forecasting has always been about
extrapolating patterns from the past to the future. Without this endeavour our
everyday routines would collapse, and all efforts to be prepared for troubles—
like earthquakes, flooding, financial crises, wars, and epidemics—would be
pointless were this not the case. Over the past decades, radical changes in
data processing speeds and data storage capacities have made a huge difference and any celebration of these developments has become accompanied by
disapproval.

The endless and often random gathering of big data has quickly found
multiple deployments in science, business, and governance. Much of this is
related to new forms of surveillance facilitated by digital metrics. Most strik-
ingly, though, much of this is powerfully driven by market mechanisms and the
financialisation of capitalism. Customer profiling, crime prevention, migration control, infectious disease control, platform economies, automated stock
exchange trading, market prognoses, and financial tools—all work with big
digital data, machine learning, and artificial intelligence, to identify patterns
and prognosticate developments that trigger automated or human interventions. The very distinctions between basic science, applied science, governance, and market mechanisms have lost much of their clarity. This new state
of affairs has huge implications not only for relations between the state, business, and civil society, but also for the unequal relations between rich and poor
countries around the world.

A politically important and remarkably centralised programme, defined
and run by the United Nations, is called the Sustainable Development Goals
(SDGs) and has the stated aim of overcoming inequalities between rich and
poor countries. Unlike its predecessor, the Millennium Development Goals
(MDGs), the new programme and its seventeen SDGs are fully metricised. Not
only do they set targets for seventeen key areas of intervention, but they also
define the precise measurements that will be used to assess the impact of all
the myriad meliorative interventions. This means that the primary objective of
the UN programme is first to put in place the necessary infrastructure to carry
out the measurements according to the defined standards. It is no exaggera-
tion to say that the SDGs are, in fact, a vast programme of metricisation. Its
primary goal is to make things comparable across space and time.

A key problem of this aim is the simultaneity of centralising and decentralising effects of digital metrics. On the one hand, digital infrastructures are characterised by the fact that they do not depend on a single centre, but are deliberately run through decentralised infrastructures (Hecht and Edwards 2007; Edwards 2010). As mentioned earlier, this also means that while the core elements of the infrastructure may be located in the tech-centres around the world, they can be applied for local purposes in any particular part of the world. Similarly, while social media have their technological, financial, and managerial centres, they are also used for more local purposes around the globe, far from these centres (Lamoureaux et al. 2021).

On the other hand, key elements of defining equivalences, categories, and codifications are located in the centres of calculation, and powerfully shape and constrain the possibilities that remain in local contexts. The obvious examples of this kind of centralisation—such as Amazon, Alphabet (Google), Apple, Microsoft, Meta and the platforms for which they offer the background service—are notable for their enormous size, global scale, degree of monopolisation, impact on local structures, and tax evasion strategies that dwarf the powers of nation-state regulation (Srnicek 2017). The opportunities and pathways for capitalism to move beyond extraversion and promote profitable businesses that contribute to increased welfare, not only in the centres but also on the African continent, will largely depend on how the mechanisms of digital centralisation and decentralisation work (Breckenridge 2018).

Like with the SDGs, any global survey needs to handle the challenge of developing categories that can travel across various borders—infrastructural, legal, political, social, and cultural—and create commensurability across a multiplicity of relevant cases. This creates a dilemma: the survey categories need to be translated into local terms in order to accurately quantify local ideas and behaviours. To allow comparisons across borders, the categories must still refer to the same thing wherever they are used, even if the phenomenon being quantified manifests itself differently in different places. In order to understand how such categories are formed, it is essential to examine the process of their creation, in other words, the practices, templates, actors, and networks that collectively constitute the expertise to draw up such categories.

The authors of the chapters in this volume do not assume that they already know how the dynamics of centralisation and decentralisation will or should develop in the future with regard to local forms of self-determination. Rather, they understand this as an open empirical question and therefore invest their energy in gaining more detailed insights through their praxiographies.

Summarising one aspect of her pioneering study of multiple linguistic and arithmetic practices among the Yoruba people of Nigeria (Verran 2001), Helen

Verran demonstrates that even mathematics and its numbers can operate in different forms and are therefore not inherently standardised. At a similarly fundamental level, Helen Robertson argues that the mathematical models of machine learning can achieve high levels of accuracy in classifying entities, but they still do not possess—in other words, understand—the relevant concept required for classification. This insight implies that centralisation and standardisation can hardly be separated from human understanding of meaningful concepts that make sense in particular local contexts, and thus remain fluid.

On a more empirical level, René Umlauf provides a disturbing insight into the practice of training a machine model to learn classification. The work is carried out by computer science students at a university in Uganda, whose role is reduced to the simple task of training the machine to repeat standardised attributions. At a similar empirical level, Emma Park and Kevin P. Donovan offer a more optimistic insight into digital forms of centralisation that are more in line with Helen Robertson's findings. They show that local variation is still possible and more powerful than is often feared. We learn that the digital apparatus centrally designed to regulate lending through standardisation does not necessarily weaken local forms of social relations and mutual support. On the contrary, users are finding creative new ways to combine digital lending with their existing social obligations. Véra Ehrenstein elaborates how forest metrics—beyond its centrally standardised methods, technologies, and funding arrangements—still rely heavily on decentralised local practices. Jonathan Klaaren's chapter explains how the right of access to information as a form of local empowerment depends on digital forms of standardisation and centralisation. He argues that centralisation and decentralisation do not simply contradict each other but unfold dialectically in the contemporary sociopolitical order of post-apartheid South Africa

Acknowledgements

I am indebted to Faeeza Ballim and Bronwyn Kotzen for their critical and encouraging comments on earlier versions of this text. Two anonymous reviewers helped to make the argument clearer and more accessible. I am particularly grateful to Emmanuel Didier, who took the trouble to read the text carefully and to offer substantial critical advice. The remaining weaknesses in the text are, of course, my own.

Bibliography

Abdul Jameel Latif Poverty Action Lab, MIT. n.d. http://www.povertyactionlab.org/. Accessed July 08, 2023.

Arendt, Hannah. (1958) 1998. *The Human Condition*. Chicago: University of Chicago Press.

Bail, Christopher A. 2021. *Breaking the Social Media Prism: How to Make our Platforms Less Polarizing*. Princeton: Princeton University Press.

Berg, Marc. 1997. *Rationalizing Medical Work: Decision-support Techniques and Medical Practices*. Cambridge, MA: MIT Press.

Bloor, David. (1976) 1991. *Knowledge and Social Imagery*. London: University of Chicago Press.

Boltanski, Luc. 2011. *On Critique: A Sociology of Emancipation*. Cambridge: Polity Press.

Bowker, Geoffrey and Susan Leigh Star. 1999. *Sorting Things Out: Classification and its Consequences*. Cambridge, MA: MIT Press.

Breckenridge, Keith. 2014. *Biometric State: The Global Politics of Identification and Surveillance in South Africa, 1850 to the Present*. Cambridge: Cambridge University Press.

Breckenridge, Keith. 2018. "The Global Ambitions of the Biometric Anti-bank: Net1, Lockin and the Technologies of African Financialisation." *International Review of Applied Economics* 33, no. 1: 93–118.

Bruno, Isabelle and Emmanuel Didier. 2013. *Benchmarking: l'État sous pression statistique*. Paris: Éditions La Découverte.

Bruno, Isabelle, Emmanuel Didier, and Julien Prévieux. 2014. *Statactivisme Comment Lutter Aavec des Nombres*. Paris: Zones.

Brunsson, Nils and Bengt Jacobsson, eds. 2000. *A World of Standards*. Oxford: Oxford University Press.

Callon, Michel, Yuval Millo, and Fabian Muniesa. 2007. *Market Devices*. Sussex: Wiley Blackwell.

Carruthers, Bruce G. and Wendy Nelson Espeland. 1991. "Accounting for Rationality: Double-Entry Bookkeeping and the Rhetoric of Economic Rationality." *American Journal of Sociology* 97, no.1: 31–69.

Daston, Lorrain and Peter Galison. 2007. *Objectivity*. New York: Zone Books.

Desrosières, Alain. 1998. *The Politics of Large Numbers. A History of Statistical Reasoning*. Cambridge, MA: Harvard University Press.

Didier, Emmanuel. 2020. *America by the Numbers: Quantification, Democracy, and the Birth of National Statistics*. Cambridge, MA: MIT Press.

Didier, Emmanuel. 2021. *Quantitative Marbling*. Anton Wilhelm Amo Lectures 7, edited by Matthias Kaufmann, Richard Rottenburg and Reinhold Sackmann. Halle (Saale): Martin-Luther University.

Edwards, Paul N. 2010. *A Vast Machine: Computer Models, Climate Data, and the Politics of Global Warming*. Cambridge, MA: MIT Press.

Espeland, Wendy Nelson and Mitchell L. Stevens. 1998. "Commensuration as Social Process." *Annual Review of Sociology* 24: 313–43.

Espeland, Wendy Nelson and Berit Irene Vannebo. 2007. "Accountability, Quantification and Law." *Annual Review of Law and Social Science* 3, no. 1: 21–43.

Espeland, Wendy Nelson and Michael Sauder. 2007. "Rankings and Reactivity: How Public Measures Recreate Social Worlds." *American Journal of Sociology* 113, no. 1: 1–40.

Espeland, Wendy Nelson and Michael Sauder. 2016. *Engines of Anxiety: Academic Rankings, Reputation, and Accountability*. New York: Russell Sage Foundation.

European Centre for Disease Prevention and Control (ECDC). n.d. https://www.ecdc.europa.eu/en. Accessed July 08, 2023.

Extreme Citizen Science Blog. n.d. https://uclexcites.blog. Accessed July 08, 2023.

Ezrahi, Yaron. 1990. *The Descent of Icarus: Science and the Transformation of Contemporary Democracy*. Cambridge, MA: Harvard University Press.

Foucault, Michel. (1966) 1973. *The Order of Things: An Archaeology of the Human Sciences*. New York: Vintage Books.

Foucault, Michel. (2004) 2006. *Sicherheit, Territorium, Bevölkerung: Vorlesung am Collège de France, 1977–1978*. Geschichte der Gouvernementalität I. Frankfurt am Main: Suhrkamp.

Future of Life Institute. 2023. Pause Giants AI Experiments: An Open Letter. https://futureoflife.org/open-letter/pause-giant-ai-experiments/.

Future of Life Institute. (2004) 2006a. *Die Geburt der Biopolitik: Vorlesung am Collège de France, 1978–1979*. *Geschichte der Gouvernementalität II*. Frankfurt am Main: Suhrkamp.

Garfinkel, Harold. (1967) 2012. *Studies in Ethnomethodology*. Englewood Cliffs: Prentice-Hall.

Geertz, Clifford. (1973) 2017. *The Interpretation of Cultures: Selected Essays*. New York: Basic Books.

Guter-Sandu, Andrei and Andrea Mennicken. 2022. "Quantification = Economization? Numbers, Ratings and Rankings in the Prison Service of England and Wales." In *The New Politics of Numbers: Utopia, Evidence and Democracy*, edited by Andrea Mennicken and Robert Salais, 307–36. Cham: Springer International Publishing.

Hacking, Ian. 1975. *The Emergence of Probability. A Philosophical Study of Early Ideas About Probability Induction and Statistical Inference*. Cambridge: Cambridge University Press.

Hacking, Ian. 1990. *The Taming of Chance*. Cambridge: Cambridge University Press.

Hecht, Gabrielle and Paul N. Edwards. 2007. *The Technopolitics of Cold War: Toward a Transregional Perspective*. Washington, D.C.: American Historical Association.

Heintz, Bettina. 2010. "Numerische Differenz. Überlegungen zu einer Soziologie des (quantitativen) Vergleichs." *Zeitschrift für Soziologie* 39, no. 3: 162–81.

Hetherington, Kregg. 2011. *Guerrilla Auditors: The Politics of Transparency in Neoliberal Paraguay*. Durham, NC: Duke University Press.

Hopwood, Anthony and Peter Miller. 1994. *Accounting as Social and Institutional Practice*. Cambridge: Cambridge University Press.

Husserl, Edmund. (1934–37) 1970. *The Crisis of European Sciences and Transcendental Phenomenology. An Introduction to Phenomenological Philosophy*. Evanston: Northwestern University Press.

Jasanoff, Sheila. 1994. *The Fifth Branch: Science Advisers as Policymakers*. Cambridge, MA: Harvard University Press.

Jasanoff, Sheila. 2020. "Ours Is the Earth: Science and Human History in the Anthropocene." *Journal of the Philosophy of History* 14, no. 3: 337–58.

Jerven, Morten, ed. 2015. *Measuring African Development: Past and Present*. New York: Routledge.

Joas, Hans. 1996. *The Creativity of Action*. Chicago: The University of Chicago Press.

Jonas, Hans. 1979. *Das Prinzip Verantwortung. Versuch einer Ethik für die technologische Zivilisation. Suhrkamp*. Frankfurt am Main.

Krüger, Lorenz, Lorraine J. Daston, and Michael Heidelberger, eds. 1987. *The Probabilistic Revolution*, Cambridge, MA: MIT Press.

Lamoureaux, Siri, Enrico Ille, Amal Hassan Fadlalla, and Timm Sureau. 2021. "What Makes a Revolution "Real"? A Discussion on Social Media and Al-Thawra in Sudan." In *Digital Imaginaries. African Positions Beyond the Binary*, edited by Richard Rottenburg, Oulimata Guye, Julien McHardy and Phillip Ziegler, 124–45. Bielefeld, Berlin: Kerber.

MacKenzie, Donald A., Fabian Muniesa, and Lucia Siu. 2007. *Do Economists Make Markets? On the Performativity of Economics*. Princeton: Princeton University Press.

Mayer-Schönberger, Viktor and Kenneth Cukier. 2013. *Big Data: A Revolution that will Transform How we Live, Work, and Think*. Boston: Houghton Mifflin Harcourt.

Mennicken, Andrea and Salais, Robert, eds. 2022. *The New Politics of Numbers: Utopia, Evidence and Democracy*. London: Palgrave Macmillan.

Mennicken, Andrea and Espeland, Wendy N. 2019. "What's New with Numbers? Sociological Approaches to the Study of Quantification." *Annual Review of Sociology* 45, no. 1: 223–45.

Merry, Sally Engle. 2006. "Transnational Human Rights and Local Activism: Mapping the Middle." *American Anthropologist* 108, no. 1: 38–51.

Merry, Sally Engle. 2011. "Measuring the World: Indicators, Human Rights, and Global Governance." *Current Anthropology* 52, no. S3: 83–95.

Mervis, Carolyn B. and Eleanor Rosch. 1981. "Categorization of Natural Objects." *Annual Review of Psychology* 32, no. 1: 89–115.

Miller, Peter, Trevor Hopper, and Richard Laughlin. 1991. "The New Accounting History: An Introduction." *Accounting, Organizations and Society* 16, no. 5–6: 395–403.

Miller, Peter. 2001. "Governing by Numbers: Why Calculative Practices Matter." *Social Research* 68, no. 2: 379–96.

Morgan, Mary S. 2012. *The World in the Model: How Economists Work and Think*. Cambridge: Cambridge University Press.

Morgan, Mary S . 2022. "Narrative: A General-Purpose Technology for Science." In Narrative Science Reasoning, Representing and Knowing Since 1800, edited by Mary S. Morgan, Kim M. Hajek, and Dominic J. Berry, 3–30. Cambridge: Cambridge University Press.

Mugler, Johanna. 2018. *Measuring Justice: Quantitative Accountability and the National Prosecuting Authority in South Africa*. Cambridge: Cambridge University Press.

National Oceanic and Atmospheric Administration (NOAA). n.d. http://www.noaa .gov/ Accessed July 08, 2023.

Nietzsche, Friedrich. (1887) 2006. *On the Genealogy of Morals*. Cambridge: Cambridge University Press.

Nyabola, Nanjala. 2018. *Digital Democracy, Analogue Politics: How the Internet Era is Transforming Kenya*. London: Zed Books.

Osborne, Thomas and Nikolas Rose. 2003. "Do the Social Sciences Create Phenomena? The Example of Public Opinion Research." *The British Journal of Sociology* 50, no. 3: 367–96.

Piketty, Thomas. 2014. *Capital in the Twenty-first Century*. Cambridge, MA: The Belknap Press of Harvard University Press.

Poovey, Mary. 1998. *A History of the Modern Fact: Problems of Knowledge in the Sciences of Wealth and Society*. Chicago: University of Chicago Press.

Porter, Theodore. 1986. *The Rise of Statistical Thinking 1820–1900*. Princeton: Princeton University Press.

Porter, Theodore. 1995. *Trust in Numbers. The Pursuit of Objectivity in Science and Public Life*. Princeton: Princeton University Press.

Porter, Theodore. 2012. "Thin Description: Surface and Depth in Science and Science Studies." *Osiris* 27, no. 1: 209–26.

Power, Michael. 1997. *The Audit Society: Rituals of Verification*. Oxford: Oxford University Press.

Rottenburg, Richard. 2009. "Social and Public Experiments and New Figurations of Science and Politics in Postcolonial Africa." *Postcolonial Studies* 12, no. 4: 423–40.

Shapin, Steven and Simon Schaffer. 1985. *Leviathan and the Air-pump. Hobbes, Boyle and the Experimental Life*. Princeton: Princeton University Press.

Srnicek, Nick. 2017. *Platform capitalism*. Cambridge: Polity Press.

Verran, Helen. 2001. *Science and an African Logic*. Chicago & London: The University of Chicago Press.

Weber, Max. (1904)1972 . "Die Protestantische Wirtschaftsethik und der Geist des Kapi-
talismus." In *Gesammelte Aufsätze zur Religionssoziologie*. Band I, 17–206. Tübingen:
Mohr.

Weber, Max. (1904)1973. "Die 'Objektivität' Sozialwissenschaftlicher und Sozialpo-
litischer Erkenntnis." In *Gesammelte Aufsätze zur Wissenschaftlehre*, 146–214. Tübingen: Mohr.

CHAPTER 2

Weighing the Forest: Field Measurements, Remote Sensing, and Carbon Payments in Central Africa

Véra Ehrenstein

1   Introduction

A thirty-minute drive from Libreville, the road is bordered by a dense forest on both sides. The taxi stops next to a few parked cars and a wooden cabin where a young woman greets me. She is an eco-guide wearing the camouflage uniform of the Gabonese state agency in charge of the national parks. As I agree to a guided tour, we first pause in front of a board, the colours of which are fading in the humid climate. An image of André Raponda-Walker (1871–1968) welcomes us to the "wood of giants."[1] "He was the first Gabonese botanist, and also a clergyman, the son of an English merchant and a Mpongwe princess,"[2] explains the guide, giving me a glimpse of the country's complicated history. As we enter the forest, she points at a tall tree on our right: "a centennial *Okoume* tree is saluting us." A subendemic species to the Gabonese rainforest, *Aucoumea klaineana* constitutes 80 percent of the trees in the Arboretum Raponda-Walker. The trade of *Okoume* timber to supply Europe with plywood was a major source of revenues in the booming decades of the logging industry, from the 1940s and the end of the French colonial empire, to the 1970s and the rise of oil as a highly profitable resource in the postcolonial nation (Pourtier 1989, 147). The coastal forests of the Arboretum were logged in the late nineteenth century before being granted protected status in the 1950s (Walters et al. 2015). Along sandy footpaths criss-crossed by a web of surface roots, the largest, most spectacular trees are accompanied by a wooden sign displaying their vernacular and scientific names—and the logo of a sponsor, the French oil company Total. Except for the presence of taxonomy, the place bears little resemblance to European imperial gardens and their curated specimens. The Arboretum on the shores of

---

1   Arboterum, Libreville, August 25, 2019.
2   On the communities living around present-day Libreville and the trading relations with Europeans in the 19th Century, see M'Bokolo (1981).

FIGURE 2.1   A Sunday morning in the Arboretum Raponda Walker near Libreville, Gabon
              PHOTOGRAPH: BY THE AUTHOR

the Altantic feels like a real rainforest;[3] a hospitable forest though, which teenagers, joggers, families, and dog walkers have, like myself, come to visit on this Sunday morning of the long dry season (Figure 2.1).

Why open the chapter in this forest? A clue is nailed to the bark of a tree. A metal tag indicates that, besides welcoming urban dwellers for walks in nature, the forest is also an object of study. One reason why tropical forests, like this one, have come to be of great interest to scientists and policy-makers alike, is the quantity of carbon contained in the woody tissues of the trees, and that is the topic of this chapter.

---

3   The liana-dense areas of the forest were chosen for background filming in the 2016 Hollywood-produced *The Legend of Tarzan*.

2   Carbon Quantification

The climate crisis has put forests in the spotlight. Plants use sunlight and water to convert carbon dioxide into sugars through photosynthesis, allowing them to live, grow, and build woody structures.[4] Large datasets and global models indicate that forestlands sequester a considerable proportion of the carbon dioxide that is released into the atmosphere by the combustion of fossil fuels (Pan et al. 2021; Walker et al. 2021). Although the greenhouse gas is emitted unevenly across the planet, it circulates, and some of it gets absorbed in the oceans and the vegetation, where it can stay for centuries or more. Quantifying the global carbon cycle in order to investigate its effect on climate change is an ongoing international scientific effort. In particular, scientists are trying to better estimate the carbon stored in the world's forests. To do so, they rely on a suite of instruments and methods, such as networks of forest plots and measuring devices, experimental manipulations of ecosystems, and various kinds of computer models and remote sensing datasets.

In Central Africa, a region home to the second largest tropical forest biome after the Amazon, Gabon with its small, mostly urban population, and an economy fuelled by oil exports and relying on food imports, has attracted researchers interested in tropical ecosystems: forests cover 88 percent of the national territory, spreading right from the outskirts of the capital city. Across this territory, field sites, research stations, and networks of one-hectare plots delineated by discreet signs in situ (e.g., metallic tags) have been set up to study, measure, and monitor the forests. In any given forest plot, all trees are usually counted, identified at the species level where possible, and their diameter, sometimes also their height, measured. These measurements are the basis for estimating the carbon mass of a forest, equivalent here to the aboveground carbon mass of the trees. Data from Gabon's forests collected in such sites are expected to contribute to the scientific understanding of carbon stocks in the tropics, and what it implies for the future of our planet. But, as we will see, forest measurements are equally central to climate policy.

The changes that the global climate is undergoing is, indeed, a matter of international concern, urging policy makers around the world to take action and limit the rise of the atmospheric concentration of carbon dioxide. In the mid 2000s, at the United Nations (UN) climate talks, the following idea started being discussed: developed countries could reward developing countries for reducing the carbon losses incurred from deforestation and forest

---

4   Some of that carbon is re-emitted through respiration.

degradation, and maybe in exchange obtain carbon credits to offset their own emissions.[5] The policy framework known as REDD+ quickly moved away from the global market mechanism that was initially envisioned and evolved into a constellation of pledges and protests, national strategies, and offsetting projects (Turnhout et al. 2017; Ehrenstein 2018; Asiyanbi and Lund 2020). From the outset, the Norwegian government positioned itself as a major funder of REDD+, as it sought to rely on tropical forests to achieve carbon neutrality (Hermansen and Kasa 2014). Its oil revenues have sponsored capacity building programmes managed by development institutions (the World Bank and UN agencies), smaller projects by civil society and research organisations, and bilateral agreements. Countries with high deforestation rates—like Brazil and Indonesia—were early recipients of this result-based aid, in which official development assistance funding is indexed on decreasing forest loss.[6] A decade later, in September 2019, Gabon's became the first government in Central Africa with which Norway passed a bilateral deal that would reward consistently low deforestation and increases in carbon storage.

The understanding of the earth's climate system and the commitments to prevent the loss of carbon-storing forestlands both rely on measurements, calculations, and estimates. My aim, therefore, is to talk about carbon in the Gabonese forests as an object of quantification.[7] First, the chapter casts light on the technologies and people that make the infrastructure of carbon accounting needed to translate Norway's promise into a payment. It then examines the development of a statistical model to indirectly weigh trees in tropical forests and a data campaign tailored to the needs of new space sensors. Finally, it returns to the bilateral result-based agreement and the political rhetoric of a green nation it feeds into. These four episodes each highlights a specific way in which scientific practices, technologies, funding arrangements, and politics come together to make a particular place—a highly forested country in Central Africa—have planetary significance. We will see African, European, and North American scientists looking at the forest as a global knowledge frontier and a resource to secure research funding, while governmental officials try to capitalise on the forest to find a solution to the ecological mess the world has gotten itself into.

---

5   The + was added later and stands for: the role of conservation, sustainable management of forests and enhancement of forest carbon stocks. For an overview of REDD+, see Angelsen et al. (2018).

6   Guyana, a country with low deforestation rates, was also among these early recipients (Hook 2020).

7   By scrutinising how carbon is counted and accounted for, this chapter contributes to the literature on "carbon accountability" (Gupta et al. 2013).

The main argument is that the planetary re-positioning of tropical forests as precious stores of carbon depends as much on seeing from afar, through sophisticated space sensors and complex computer models, as it does on going into these forests, which, in turn, involves setting up research sites and monitoring plots, organising arduous field missions, mobilising sponsors for it, cutting down trees, retrieving old datasets, standardising measurement protocols, and so on. All these efforts invested in weighing the forest have their own complications and the outcome is, metrologically speaking, highly uncertain.[8] And yet, there is no other way to quantify the carbon stored in hundreds of million hectares of tropical forests and account for their importance to the global climate.

The content of this chapter is based on fieldwork carried out in 2019. I conducted in-depth interviews with scientists, technical experts and government officials in Gabon, the United Kingdom (UK), and at an international scientific conference in Milan, Italy, while reading the scientific and technical literature our conversations alluded to.[9] With this research, I wish to present an illustration of how, at a time of what Achille Mbembe calls the "combustion of the world" (Mbembe 2020, 17–25), Science and Technology Studies (STS) can engage with the planetary project of mitigating climate change, by paying attention to the way it plays out in specific African contexts.[10]

3   Environmental Sciences in Africa

A good place to start, in order to position this piece within STS, is Bruno Latour's text on the "circulating reference" (Latour 1999, 24–79). In the Brazilian Amazon, the presence of identification tags (e.g., metallic tags) turns the savanna-forest landscape into a "minimalist" laboratory for soil scientists (Latour 1999, 32). By looking at the use of maps, protocols, sampling devices,

---

8   For a reflection on the meanings of uncertainty in comparable contexts, see Walford (2017) on the uncertain nature of data generated by a meteorological tower in the Brazilian Amazon and Goldstein (2022) on the tension between political and scientific uncertainties around carbon measurements in Indonesian peatlands.

9   This chapter directly builds on twenty-six semi-structured interviews, as well as observations and more informal conversations. I had planned to resume fieldwork in Gabon in 2020, but it was postponed due to the COVID-19 pandemic. As the data mobilised here is limited, I consulted researchers with relevant experience and knowledge to validate my interpretation.

10   In Mavhunga's (2017, 7) typology, this chapter belongs to the body of works that follows "the traveling [of] Western artifact, idea, or expert" to Africa.

and notebooks, Latour traces the translations of matter into form, through which "truth-value circulates" (Latour 1999, 69)—from the field plot, to university office, first in Manaus, then in Paris, to facts and figures in academic writings. The abstraction process makes the forest intelligible in new ways. As Helen Verran suggests, this is how ecological properties are established, as "real, abstract entities" that are "immanent in land areas and transcend them" (Verran 2002, 749). The biomass of a forest in tonnes of carbon per hectare is an ecological property of great interest that is calculated based on field measurements from particular locations and associated with more generic climatic conditions, soil characteristics, and vegetation types. In Verran's terms, carbon is an "enumerated materiality" (Verran 2010, 171): material, because carbon elements are essential to life on earth, enumerated, because the objective, here, is to quantify how much is stored in the woody tissues of trees.

As soil samples are collected, or trees measured, the rest of the forest disappears, but this de-contextualisation does not imply that the field-as-lab is disconnected from everything else. Science, Latour further suggests, takes place in context within a "circulatory system" (Latour 1999, 80): standardisation of measurements, classifications, theories, methods, skills, instruments, but also political interests, infrastructures, and financial supports, are necessary to sustain scientific activities. We can turn to Paul Edwards' (2010) history of climate modelling to see that the "vast machinery" of knowledge production about the habitability of our planet has built upon, among other things, the standardisation of meteorological observations in the age of empires, the possibility of a nuclear winter during the Cold War, and politicians in the United States deciding to fund more research instead of taking action. The object of this chapter—the carbon mass of tropical forests—belongs to that history, as a component of the Earth System that, in the last two decades, scientists have sought to integrate into their models using a variety of data, including from space sensors.[11] But due to the immanence of forests in specific land areas, quantifying how much carbon the trees contain immediately raises particular context-related issues that climate modelling does not.

One issue that jumps to mind is that forests are most often inhabited. Knowledge and ignorance of vegetation dynamics inform decisions about how and by whom forestlands ought to be used, which may affect people's lives and livelihoods (Fairheard and Leach 1995).[12] This chapter, however, is concerned with a different type of context-related issue, namely the transnational

---

11   On the emergence of global ecology in response to climate change, see Kwa (2005).

12   Many case studies now document the local effects of monetising forest carbon for REDD+.
        To cite just one example, see Asiyanbi (2016).

collaborations through which tropical forests are quantified. Although her topic is genomics, Iruka Okeke diagnoses the crux of the matter when she argues that uneven access to financial resources "pushes would-be collaborative partners into positions allied to givers and receivers of aid" (Okeke 2016, 462). African scientists frequently end up being specimen collectors for their foreign colleagues, whose "'postal' and 'parachute' research inevitably addresses remote or 'global' questions, not local ones" (Okeke 2016, 462). Material inequalities, then, translate into unequal distribution of research leadership, geographical mobility, and scientific credits (Crane 2013; Geissler 2013). While the situation is a direct outcome of recent macroeconomic development policies (Rottenburg 2009; Tousignant 2018), these asymmetries are also a reminder of the enduring legacy of imperial knowledge production about Africa, and in the STS scholarship on botanical knowledge, the scramble for healing plants illustrates that well (Osseo-Asare 2014; Langwick 2021; on biodiversity science more broadly, see Asase et al. 2021). Carbon quantification, one might argue, is yet another example of *extraverted* research, where research design, data processing, and theoretical interpretation are done elsewhere (Hountondji 1990).

When it comes to tropical forests and carbon, international collaborations running the risk of turning into parachute research are not restricted to Africa. In her analysis of an international research programme (the Large-Scale Biosphere-Atmosphere Experiment in Amazonia), Myanna Lahsen suggests that Brazilian policy makers experienced the "Northern dominance of science" as a threat to their control over the narrative about deforestation in the Amazon (Lahsen 2009, 362). All the while, the interest of European and US research funding agencies fluctuated depending on whether the data suggested the forests acted as a carbon sink or source (see also Fearnside 2009). Antonia Walford (2012), speaking about the same research programme, unpacks the micropolitics of fieldwork, showing that Brazilian scientists and technicians felt at times sidelined by foreign researchers appropriating the data obtained with their help. These examples show that trying to produce field-based environmental knowledge across such uneven terrains raises difficult context-related questions, such as scientific imperialism and territorial sovereignty.[13]

The carbon stored in tropical forests, in Central Africa and elsewhere, is hardly just a scientific research interest. In Gabon, public discourses depict a green nation with low deforestation rates and precious stores of carbon, that is simultaneously home to illegal logging activities and wide social disparities.

---

13   Beyond carbon quantification, the sense of a loss of sovereignty in relation to forest expertise and policy reform is a major concern among governmental officials in Central Africa. For an illustration in Cameroun, see Ongolo and Karsenty (2015).

This echoes the "dystopian paradigm" pointed out by Florence Bernault and Joseph Tonda: "the critique of a small authoritarian and corrupt dictatorship coexists with the praise of a political and economic peace haven. The environmental dream of a future equatorial Costa Rica is undermined by the nightmare of a 'Heart of Darkness' new style" (Bernault and Tonda 2009, 10).[14] Bernault and Tonda further note that Gabon is frequently "picked as a 'case' for the study of French neo-imperialism in Africa" (Bernault and Tonda 2009, 14).[15] To STS scholars, Gabon is thus known as the location of Gabrielle Hecht's (2018) "African Anthropocene." Seeking "a means of holding *the planet* and *a place on the planet on the same analytic plane*" (Hecht 2018, 112), Hecht traces the "interscalar" travel of radioactive uranium found in Gabon, across space and bodies, from deep time to an all too human history. The journey brings forth the toxic residues left behind by the exploitation of the planet's geology and the predatory relations of the *Françafrique*. The case study presented in this chapter conveys a different message than the one from the uranium mines: there, the earth has been turned inside out;[16] here, trees take the outside world in, soaking up the carbonised residues of anthropogenic excess, under the attentive gaze of scientists, experts, and policy-makers, in a brittle political context.

As I will now describe different efforts to quantify forest carbon stocks in Gabon, I propose to examine the interplay between seemingly trivial issues (disbursement procedures for field missions), political strategies (the environmental diplomacy of an autocratic regime), technical achievements (new space sensors to study the earth), and scientific discussions (around the validity of a unique equation to weigh tropical forests), in order to highlight both the short-term contingencies and long-term legacies through which particular forests may be seen to acquire planetary significance (which they can also quickly lose).

---

14   Author's translation. As Bernault and Tonda (2009) suggest, Gabon is economically prosperous (compared to other countries in the region) due to its oil reserves and politically stable mainly because it has been ruled by one family since 1967. My understanding of the reference to Joseph Conrad's novel *Heart of Darkness* set in the colonial Belgian Congo, is that it alludes to the dehumanising racist image of Central Africa the novel conveys and maybe also to current problems in Gabon around the now illegal ivory trade, hence the contrast with the "environmental dream."

15   Author's translation.

16   "Inside-out Earth" was the title of Gabrielle Hecht's lecture at University College London, October 23, 2019.

4   A National Carbon Accounting Infrastructure

The Gabon-Norway deal announced in 2019 anticipated that carbon gains and losses could be estimated across tens of millions of hectares. The letter of intent stated that the African nation shall be rewarded for reducing emissions from deforestation and forest degradation (carbon losses from cutting down trees) and increasing the "removals from land remaining forest land" (carbon gains in re-growing forests) (CAFI 2019). Up to 150 million dollars could be transferred over ten years, starting retrospectively in 2016. The payment would be disbursed at an estimated rate of ten dollars per tonne of carbon dioxide. The Gabonese Space Observation Agency and the National Resource Inventory of Gabon Parks Agency were essential to the implementation of such an agreement. Since 2018, the two institutions were being supported by a programme called the Central African Forest Initiative, to which Norway was a major donor. Gabon had secured a grant from it to establish a carbon accounting infrastructure centered on field plot measurements and remote sensing data analysis.

The creation of earth observation facilities had been made financially possible through a debt cancellation agreement with France. Instead of paying the debt back, the Gabonese government committed fifty million euros to environmental and forestry programmes, of which ten million euros was to help create the Space Observation Agency. The French and Gabonese Presidents announced the debt swap in 2007 as they walked in the Arboretum near Libreville during a diplomatic visit epitomising the *Françafrique* (Bernard and Jakubyszyn 2008; Soir 3 Journal 2007).[17] Twelve years later, the agency "is entirely managed by Gabonese," emphasised its director on our way to the facilities twenty kilometres from the capital city.[18] For him, "the biggest challenge was to train human resources, no one had professional skills in remote sensing in Gabon five years ago." As I was shown around, and observed a huge antenna detect a nearby satellite (Figure 2.2), I met with engineers who received state scholarships to study abroad and came back with the required skills. They were busy interpreting archives of satellite images retrieved from American (Landsat) and European (Sentinel) optical sensors in order to map the evolution of land cover across the national territory. Datasets for 1990, 2000, 2005, 2010, and 2015 had already been analysed and independently validated. This work would be key to calculating the results Norway pledged to pay for (CNC

---

17   The French President's trip to Libreville followed his visit to Senegal where he gave an infamous speech marked by colonial paternalism and racialised prejudices (Mbembe 2007).

18   Observations at the Space Observation Agency, Nkok, August 29, 2019.

FIGURE 2.2
A ground antenna of the Gabonese
agency for Earth observation, Nkok,
Gabon
PHOTOGRAPH: BY THE AUTHOR

2020). The accounting method consisted of, first, imposing a grid of sampling units upon the territory. For every annual dataset, each unit was assigned a land use category (e.g., forest land,[19] cropland, grassland, settlements, or other) and, in case of forest land, a subcategory related to the type (e.g., primary forests or secondary forests). Matrices were, then, to be generated to track what happened to the units from one period to another (e.g., between 1990–2000 and 2000–2005). This revealed which area experienced a change, such as switching from forest land to cropland, which would be an instance of deforestation, or from forest land/primary forests to forest land/secondary forests, which would be an instance of degradation. The aim was to quantify square kilometres of forest land lost and gained over time and the task was not always easy. I was told, for example, that due to the cloudiness of the African equatorial climate, almost no exploitable image could be found for some parts of the territory.

To convert area estimates into tonnes of carbon, field measurements and a different set of expertise and tools were needed (how exactly a carbon stock is obtained is discussed later). Providing such data was the job of the National Resource Inventory, a division of Gabon Parks Agency. The division was created

---

19   The forest is here defined as an area of at least one hectare covered by at least thirty percent of trees that are at least five meters high.

in 2012 together with a national network of 104 one-hectare plots. A first census of the network was done between 2012 and 2014, and as it was deemed to be of scientific quality, the data have since been used in peer-reviewed publications (e.g., Poulsen et al. 2020). This first census was funded by a US interagency technical cooperation programme, the Food and Agriculture Organization and an Indian-Singaporean agrobusiness company called Olam. Olam has entered several public-private partnerships with the Gabonese government, including a few palm oil plantations (on Olam's economic importance see Mouissi 2018). My understanding is that a new sustainability criterion in the commodity's supply chain—the protection of high carbon stock forests when establishing new plantations—had motivated its support of a first carbon inventory. Data collected across the plot network were used to estimate a national average carbon stock and helped to set the upper limit of the high carbon stock threshold at 118 tonnes per hectare (Burton et al. 2007, 304). Below this value, existing vegetation may be replaced by oil palms, the production of which still qualified as sustainable. High carbon stock assessments had been pioneered in Southeast Asia, in response to environmental organisations and their zero-deforestation campaigns shaming global food brands for destroying rainforests (Cheyns et al. 2019). The global governance of palm oil indirectly made the first quantification of Gabon's carbon stocks financially possible, which revealed that its forests had large and heavy trees.

The inventory network was established following a sampling strategy in order to capture enough variations across the forest lands—one plot, for example, ended up in the *Okoume*-rich secondary coastal forest near Libreville (Poulsen et al. 2020, 3).[20] For the purpose of carbon accounting, the 104 sites were classified by disturbance type: primary, secondary, and logged forests. These categories, and the corresponding field measurements, provided data to translate land-use change matrices into carbon losses and gains (CNC 2020). Carbon is lost if some units switch from, for example, forest land/primary forests to a non-forest category (e.g., settlement). To quantify the loss, the affected area size is multiplied by an emissions factor, which is an average carbon stock value per hectare derived from the subset of plots located in undisturbed (primary) forests. If, instead, the area is converted into secondary forests, another emission factor is used that is equivalent to the difference between the average carbon mass of primary forests and that of secondary forests (the latter obtained from the subset of plots in secondary forests). But if the forest remains untouched, its trees might accumulate more carbon. To quantify this

---

20   The National Resource Inventory was intending to increase the number of plots to five
      hundred.

potential gain, the area size must be multiplied by a removal factor equivalent to an average rate of carbon sequestration in primary or secondary forests depending on the forest type. Calculating a rate requires comparing carbon stocks at different points in time. The dataset of the first census did not provide such a dynamic vision of the forests, and that is why a new survey was launched in 2019.

Conducting a forest inventory in dense tropical vegetation is no mean feat. Field agents travel from Libreville in small groups to distant locations, where they measure trees following standard protocols developed within the international scientific community. A botanist laconically summarised, "Some plots are far from the road, you might walk several days in the forest and camp there. Sometimes there are waterways. It's hard."[21] Inventory teams usually visit nearby villages and hire volunteers to guide them through the forest. The work is dangerous (malaria, snakes, elephants, poachers, accidents) and costly. Cars, fuel, cellular phones, GPS, measuring devices, tents, gears, medicines, food, and cash for the helpers must be provided for. The operation of the National Resource Inventory, I was told, remained contingent on intermittent external funding and its workforce in a state of financial precarity.

In 2019, the Central African Forests Initiative was sponsoring the new census and oversight of the project had been delegated to the French Development Agency due to its long-standing presence in Gabon—a postcolonial legacy. The disbursement of tangible local currency bills was to follow strict procedures. Even buying food cans for field missions proved a hassle, involving quotes from a wholesaler, approval by the development institution, and payments by cheque. The administrator of the National Resource Inventory concluded that: "You can be sure I'm back in the store and I'm told they don't have the cans anymore, or not enough, or a different type at a different price."[22] Before occupying this position, she was the head of cashiers in a big supermarket and, therefore, was quite certain that her small orders would never be the priority for a wholesaler. For the new field missions, as in the first census, she simply went to a supermarket, kept the receipts and filled in forms. These had been sent back full of "ineligibilities." As it was the start of the project, an "operational incident" for non-compliance had not been raised yet. Retaining pledged money is not a rare thing to do for development agencies. At the French Development Agency's office in Libreville, I learned that most of its forest-related programmes (e.g., on logging legality) had rather low rates of

---

21   Interview 33, Botanist, Libreville, September 2, 2019.
22   Interview 35, National Resource Inventory Administrator, Libreville, September 4, 2019.

disbursement due to suspicions around misappropriation of funds.[23] But from the perspective of the National Resource Inventory, the disbursement procedures may just end up impeding time-sensitive and scientifically valuable field measurements.

Carbon accounting requires tracking changes in land areas over time and assigning them a value as loss or gain. The method, here, relied on remote sensing techniques and plot inventories, for which an adequate infrastructure and skillset had to be in place. To secure funding for it, two governmental agencies sought to mobilise multiple resources, while remaining dependent on external partners controlling the funds, a situation also shared by Gabonese scientists working in the field of tropical ecology, as we shall see now.

5   One Equation to Weigh any Tropical Forest

Field data collected during a forest inventory mainly consist in measurements of the diameter, and sometimes the height, of all the trees within a plot (except the tiny ones). Yet, the variable of interest to the Gabonese government, its Norwegian sponsors and the scientific community studying the carbon uptake of tropical forests, is the aboveground dry biomass of the trees (half of that is carbon). Weighing the forest poses an ecological-metrological dilemma as the most direct method requires harvesting trees. Trees are uprooted, cut into pieces, which are weighed on a scale brought into the forest, while samples of fresh wood are taken to a lab, dried, and weighed as well.[24] The method is costly, extremely time consuming, and destructive. This is why an indirect method based on non-destructive measurements is used. But to translate diameter and height values into carbon stocks, some trees have to be sacrificed. In the 2010s, the Gabonese Research Institute for Tropical Ecology participated in two of such destructive sampling projects that aimed to improve some of the tools available to scientists for estimating the biomass of tropical forests.

Estimating the biomass of a forest depends on statistical techniques. Destructive sampling generates data suitable for establishing a relation, an allometric equation, between three (or four) variables: the first variable is biomass, while the two (or three) others are predictors and include one (or two) structural feature of the trees, their diameter (and sometimes the height), and a species-specific trait called wood density. In Gabon, once the first census of the plot network was completed, an allometric equation known as the

---

23   Interview 29, Junior Project Manager, Libreville, August 29, 2019.

24   For a similar dilemma in the measurement of body composition, see O'Connell's (1993).

"pantropical model" was used to translate the measurements into mass estimates (Poulsen et al. 2020, 4). This equation is meant for large-scale assessment of multispecies forests. It seeks to facilitate the quantification of carbon stocks from non-destructive inventories across the tropics, so that estimates from different sites may be aggregated.

The pantropical model was developed through an international scientific collaboration initiated in the early 2000s. French ecologists with experience of working in French Guiana, together with colleagues from the United States and Japan, started compiling structural and mass measurements from published destructive sampling studies carried out in tropical forests since the 1950s (Viard-Crétat 2015, 344–46). Data from two thousand four hundred weighed trees were used to develop a first set of equations published in 2005 (Chave et al. 2005). But the geographical coverage of that dataset was limited: there were no trees from Africa. In response to this, funding became available to conduct destructive sampling studies on the continent. In Gabon, the Research Institute for Tropical Forestry channelled overseas aid from the European Commission to weigh one hundred trees from a logging concession in the northeast of the country (Ngomanda et al. 2014). As similar studies were done across the continent, an updated dataset was eventually published in 2014. It included four thousand sacrificed trees from fifty-eight sites—one in Gabon—based on which revised pantropical models were obtained (Chave et al. 2014).[25]

In addition to allometric relations, another global dataset was created around the same time, this one compiling wood density values, which are needed to convert volume into mass (Chave et al. 2009). Here, data for African tree species were easier to find. Some, for example, can be traced back to a 1955 report of the French Technical Centre for Tropical Forestry (Sallenave 1955). The report detailed the physical and mechanical properties of samples collected across the colonies in order to promote the commercial uses of these tropical woods. The work underpinning the report had begun after the first world war, as overexploited forests in metropolitan France drew attention to overseas territories. Before its independence in 1960, Gabon was a major source of timber and also a site of experimental forestry for its colonisers.[26] One can, therefore, speculate that some of the wood used to quantify the density values

---

25   Different equations are proposed depending on whether height measurements are available or not. Inventory data in dense tropical forests might include only tree diameters as it is easier to measure.

26   The Technical Centre for Tropical Forestry communicated its findings in a technical journal, *Bois et Forêts des Tropiques,* still published today. In the 1950s, Gabon's forests were frequently mentioned and the Mondah forest, where the Arboterum is now located, was a major research site for experimental Okoume plantations (Aubréville 1954).

found in the 1955 report had been sampled in these forests. Fifty years later, the same data is instrumental to the work of scientists trying to estimate the biomass of African forests.

In Libreville, the office building of the Gabonese Research Institute for Tropical Forestry is another legacy of the colonial project as it used to host the French Technical Centre for Tropical Forestry. I met there with a senior scientist, a paleoecologist by training, who had lived abroad, in France and Germany, before returning to Gabon. The work of his institution relied on international sponsors, through research partners in Europe, the United States, and Japan, as well as technical aid programmes. The second destructive sampling study done in Gabon to weigh another hundred trees, which he was involved in, was made possible through a regional capacity building project. The World Bank managed the funding that came from a multilateral organisation (the Global Environmental Facility), and European consultants in Yaoundé, Cameroon, supervised the implementation in the six partner countries—Cameroon, Congo, the Democratic Republic of the Congo, Equatorial Guinea, Central African Republic, and Gabon (Fayolle et al. 2018). Reflecting on the institutional montage of the project, the ecologist remarked that, in these arrangements, "financial resources dedicated to scientific work is spent on management costs," and so money is "lost in meetings," while the project could have been more innovative scientifically speaking.[27]

One innovation discussed with enthusiasm by the ecologists and remote sensing scientists I spoke with was the use of commercial laser devices. In a forest, the laser sends a beam of photons that reflects on leaves and wood and generates three-dimensional point clouds containing billions of measurements. The visualisation of the point clouds looks like a forest. A group of UK-based remote sensing scientists have been particular active in pioneering the use of terrestrial laser scanning for volume (and biomass) estimates. At their university in London, the researchers were coding algorithms to infer, from the point clouds, the woody volume of the forests they had surveyed.[28] Gabon was the first "tropical forest environment" in which they tested their laser in 2013. The instrument had since scanned ten thousand trees in Australia, the United States, Brazil, Peru, Malaysia, Ghana, and a couple more times in Gabon. The technology

---

27   Interview 36, Senior Ecologist, Libreville, September 4, 2019.
28   Interviews 1, 5, 6, and 15, Remote Sensing Scientists, London, November 15, 2018, February 7, 2019, February 19, 2019, and June 10, 2019. To convert 3D point clouds into meaningful information, algorithms are coded to extract each tree, separate the points associated with leaf material from wood material, and calculate a volume. Only 10 percent of the initial dataset may be useful.

might help to improve the pantropical allometric model mentioned earlier. The equations have the merit to exist, but due to the scarcity of destructive sampling studies, the estimates they allow to compute are highly uncertain.[29] In particular, large and heavy trees are under-represented in the global dataset, making the pantropical model poorly suited to weigh the big stores of carbon characteristics of Central African forests. Scanning a greater number of large trees might be easier, and better for the forests, than weighing them (Disney 2019), even though new sources of uncertainty (e.g., from converting point clouds into tree volume) come with the new method. Laser scanning was, nevertheless, seen as promising to advance the understanding of tree structure, and the ecologist I met in Libreville was keen to use it.

To collect data and samples in Gabon's forests, foreign researchers must enter a data sharing agreement with a national institution. In 2019, the Research Institute for Tropical Forestry was the partner of a new project co-led by the UK-based scientists, whose objective was to laser-scan trees in a forest concession in order to assess biomass (and carbon) changes associated with logging. The Gabonese institution had asked if they could also take the device to the coastal rainforest of the Arboretum and help study this puzzling eco-system. From the perspective of underfunded scientists on the government's payroll, in a context of reduced income due to the decline of global oil prices in 2014, external interests in the country's forests could prove useful to sus-tain a national scientific community and try to carve out support for their own research agenda.

6   Ecological Representatives to Calibrate Space Sensors

Gabon's forests have not only contributed dead trees to improve tools used by scientists to estimate the biomass of tropical ecosystems. These forests are also of value to the international scientific community as living ecologi-cal representatives. The enumerated materiality of carbon stored in terrestrial vegetation is an important element in the study of the earth conceptualised as an integrated system of biogeochemical cycles and energy transfers. In the last decade, space agencies such as the US National Aeronautics and Space Administration (NASA) and the European Space Agency, have invested in new satellite-borne sensors that might provide datasets better suited to the needs

---

29   Allometric relations provide precise estimates if the trees harvested and weighed to establish them are similar enough (same species, age distribution, climate, soil etc.) to the standing trees the biomass of which is quantified.

of Earth System modelling.[30] In particular, active remote sensing using laser and radar instruments can record signals related to the height of the world's forests. While passive optical technology—think of Google Earth—is great at indicating the presence of vegetation, it does not give information about the size of the trees. Active sensors may do that. But to translate their waveforms (laser) and backscattered signals (radar) into spatially explicit biomass estimates, field measurements from actual forests are needed, and this is how Gabon came to host a data campaign for ambitious space programmes supporting the study of the global climate system.

NASA's light detection and ranging sensor (a laser) called GEDI, for Global Ecosystem Dynamics Investigation, was plugged to the International Space Station in December 2018 (GEDI 2019). Six months later, updates from the mission were presented at the Living Planet Symposium in Milan. "Taking the pulse of our planet from space" was the moto of the conference on the importance of remote sensing for Earth System sciences.[31] GEDI was discussed in several talks, alongside BIOMASS, the satellite-borne radar that the European Space Agency aimed to launch in 2023. Some of these talks focused on a joint data collection campaign for GEDI and BIOMASS that took place in Gabon in 2015 and 2016. The campaign involved engineers and scientists from American, German, British, and French research institutions (Fatoyinbo et al. 2021). Planes were brought from the United States and France and equipped with instruments mimicking the future space sensors to fly over different forest types, as teams underneath trudged to field sites to measure the trees. The campaign had various purposes. One was to start developing algorithms to convert the height-related signal the sensors would record into mass estimates by using the field data as ground truth (diameter and hight measurements converted into mass via the pantropical allometric model).[32] The remote sensing scientists I spoke with were acutely aware that these biomass estimates would be uncertain but there is currently no alternative.

When I asked why Gabon was selected for the campaign, a scientist from GEDI listed the following reasons: "it's safe, so it minimised the security risk for the teams, and there's a lot of forests, high biomass, with different types of forest, biologically and structurally diverse", adding that "it's an oil economy

---

30   See Gabrys (2020) for a discussion of remote sensing and the planetary gaze.

31   Observations at the Living Planet Symposium, Milan, May 13–17, 2019.

32   Laser and radar sensors detect the forest's height as the difference between the signal bouncing back at the top of the canopy and the signal hitting the ground. To establish a relation between height and mass, plot measurements are used to estimate biomass and computational techniques relate that value to the measured height. For a discussion of the conventional nature of similar measurements, see Mallard (1998).

which is why the forest is intact, and there are infrastructures."[33] The infra-structures that greatly facilitated the foreigners' work were the National Resource Inventory's botanists and field agents and the Space Observation Agency. To the audience at the scientific conference, the researcher further justified the location by presenting Central Africa as a "data gap" to be filled.[34] The campaign was meant to generate missing information and supplement existing datasets from Europe, and North and South America (Asian forests were another "data gap"). The idea that Africa is lacking something is a cliché and this chapter has shown that, in the mid 2010s, data about carbon storage in Gabon's forests started being available (the first national census) but these date were not suitable for sensor calibration.

A promotional video was published online about the GEDI mission and the data campaign in Gabon (NASA 2016). It shows a diverse group of scientists in hiking gear guided through a forest (the Arboretum) by Gabonese field agents. Images of a plane ready to take off and dense forests seen from above alter-nate with interviews in a studio. The researchers explain that in such a hot and humid climate "life constantly regenerates" and Gabon "has some really dense tropical forest that has not really been studied, extensively, especially from a remote sensing perspective," reiterating the data gap argument (NASA 2016). They also emphasise the aim of the mission: to "balance the global car-bon budget"—quantifying how much carbon is stored in the atmosphere, the oceans and the land—and ultimately improve the understanding of the Earth System. For NASA and its team of remote sensing scientists, what mattered was to find a variety of tropical vegetations for which airborne and ground data could be obtained to support the calibration and validation of a new space sensor (Figure 2.3).

The 2015–2016 data campaign was actually not the first time that these forests were involved in remote sensing research. One of the first pantropical carbon stocks maps, which was published in 2011, was obtained with plot mea-surements from across the tropics, including in Gabon. The United States-based researchers leading the project needed field data to develop computational techniques to translate the signal recorded by a satellite-borne altimeter into a carbon map (Saatchi et al. 2011). The field measurements were obtained from research plots scattered across the continent—areas where the forest, its fauna or flora, had been studied by ecologists who, among other things, measured

---

33   Interview 13, Remote Sensing Scientist, Milan, May 16, 2019. French Guiana is another
      place where a joint campaign was carried out, see Viard-Crétat (2015) on this French over-
      seas territory used as a laboratory for tropical ecology.
34   Observations at the Living Planet Symposium, Milan, May 13–17, 2019.

FIGURE 2.3   A data campaign for new space sensors in a high forest—and low deforestation—
country with diverse ecosystems, 2019 Living Planet Symposium in Milan, Italy
PHOTOGRAPH: BY THE AUTHOR

the trees. In the mid 2000s, these independent research sites started being integrated into the African Tropical Rainforest Observation Network, a collaborative data sharing initiative replicating what had been done in the Amazon. A scientist from the UK, drawn to tropical forests because this is "where most of life is," and out of taste for "adventure," received a fellowship that gave him substantial time and money to travel in Africa and set up this network of research plots.[35] After reviewing the literature, he contacted "the owners of the plots, the people who measured them," went to visit the sites and re-surveyed them. His first stop was Ghana where plots from a British development project in the 1980s were still in place; then Nigeria, but the plots found in the literature had been destroyed; next Cameroon and Gabon, where the re-measurements worked well.[36] The objective of the Amazonian and African observation networks was to provide a platform to answer questions about the carbon uptake of old-growth rainforests in the context of an increasingly warm planet (Hubau et al. 2020).

35   Interview 9, Global Change Scientist, London, March 25, 2019.
36   For example, some research plots in Gabon were "owned" by French scientists and others by a British researcher who would later become forest minister (next section).

These measurements have served other purposes as well, like the production of pantropical carbon maps.

When the objective is to take the pulse of our planet from space, tropical forest lands are discussed as a global knowledge frontier, about which more information should be obtained. No Gabonese representatives were interviewed in NASA's promotional film, nor, as far as I was able to trace, present at the remote sensing conference in 2019 in Italy. A German scientist involved in the data campaign indirectly hinted at this absence. One of her slides displayed a photograph of a large group comprising the Euro-American teams, Gabonese state officials and younger people posing next to a plane. The photograph was taken during an outreach day and the researcher commented: "we also provided training in order not just to go and take the data".[37] It was a different photograph, but I also saw an image of one of the campaign's aircrafts in the office of the director of the Space Observation Agency.[38] The latter had negotiated access to a military air base and obtained authorisations for the flights. The remote sensing engineers I spoke with remembered the campaign well and talked fondly about the planes. As part of the outreach, data from the airborne instruments had been shared with them, although not much more followed and the engineers were unaware whether the sensors were already in orbit or not. The latest technological breakthroughs in earth observation were of little immediate relevance to their work, given that to assess land cover changes and implement the result-based agreement, they had to use archives of satellite images and time-tested methods.

Data related to forest biomass—which active sensors ought to generate— may be useful to constrain and evaluate the computer simulations underpinning the study of the earth's climate system (Herold et al. 2019, 761–66). Here, the end goal of carbon quantification is to better understand large-scale biogeochemical cycles and the troubled state of our planet. But, as carbon-absorbing forests are immanent in land areas, their fate depends on land-use decisions, and the extent to which scientific evidence is influential there is uncertain. For the director of the Space Observation Agency, in Gabon, "because the forest covers nearly 90 percent of the territory, every human activity takes place in the forest, whether you build a road or a dam, you impact the forest,"[39] hence the expectation that giving a monetary value to carbon would alter those decisions.

---

37   Observations at the Living Planet Symposium, Milan, May 13–17, 2019.
38   Observations at the Space Observation Agency, Nkok, August 29, 2019.
39   Interview 25, Space Observation Agency Director, Libreville, August 27, 2019.

7   An Aspiring Green Champion?

The result-based agreement with Norway was announced during the 2019 United Nations Climate Action Summit in New York, and Gabon's National Minister of Water, Forests, the Seas and the Environment, gave a speech about it at a side event. He emphasised that "it's almost more important to us that Norway is putting its faith and its reputation on the line alongside our own, that's almost more important than the money" (Lang 2019). Norway's offer of ten dollars per tonne of carbon dioxide avoided, double the price in its previous bilateral agreements, was applauded. To diplomats and experts, the minister described Gabon as "high forest, low deforestation" and this was presented as the outcome of development choices made by the previous and autocratic Presidents, who happen to be father and son. In Libreville, a month before the summit, a state official had told me that "Gabon wants to put forward the lack of international policy instruments suitable for countries with lots of forests."[40] Allied with Surinam and Guyana, the Gabonese delegation was pushing for the category of High Forest Cover Low Deforestation (HFLD) countries to be further recognised in climate discussions. The deal with Norway was expected to pioneer a policy attuned to HFLDs and simultaneously raise Gabon's environmental profile internationally.

What the minister considered to be forest-friendly development choices are listed in the project documents written for the Central Africa Forest Initiative (CNC 2020, 7–10; CNC 2021, 19–24). These include: a forest code making sustainable management mandatory, associated with a ban on raw logs exports to sell higher value products and the engagement of multinational logging companies in private certification; a focus on nature conservation and national parks supervised by a governmental agency tasked with valorising their fauna and flora through scientific research and eco-tourism; and finally, private investments in industrial agriculture subjected to global sustainability criteria, especially palm oil production. In addition to these measures, the documents evoke Gabon's involvement in UN negotiations. The scientific collaborations mentioned earlier (the African plot network and the pantropical carbon map) had, for example, been showcased at the 2009 climate meeting in Copenhagen, where European and American academics were in the Gabonese delegation. Yet, in the subsequent years, Gabon was much less visible in the REDD+ space. One reason was the focus on high deforestation countries; another was that the government said it wanted to engage with the global policy on its own terms,

---

40   Interview 28, Governmental Official, Nkok, August 29, 2019.

not that of development institutions' (Walters and Ece 2018). It might have also waited to see whether anything tangible came out of these (slow) discussions.[41]

From 2009, national policy making in Gabon became driven by the "Emergence" mantra. Political scientists briefly wondered (and quickly lost their illusions) whether it might announce a departure from the discretionary political regime in place since the late 1960s (Mouity 2012). They discussed, for example, the nomination of a scientist at the head of Gabon Parks Agency: the researcher was clearly competent for the job but his British nationality raised concerns (Mouity 2012, 49–50). In June 2019, the scientist became the National Minister of Water, Forests, the Seas, and the Environment. He first arrived in Gabon in the late 1980s to do fieldwork for his PhD in tropical ecology, and then worked for a US wildlife organisation to help establish new national parks, before being granted Gabonese citizenship as he took charge of the governmental agency in 2009 (Dougueli 2019). All the while, he continued to conduct research through an academic affiliation in the UK. Having a scientist who belonged to the President's close circle certainly encouraged some of the international collaborations previously examined. As a foreign researcher involved in a couple of those projects put it: he was doing fieldwork in Gabon because he had "lots of contact high up there."[42] The country, another scientist summarized, was "a nice place to work"—if things remained as they were.[43] One can reasonably assume that the nomination of a professor in ecology as forest minister boosted Gabon's green capital on the global political scene, beyond the scientific community.[44] Within the country, however, the white conversationist was less popular, especially in rural places where forest elephants regularly destroyed villagers' crops and people rightly felt disenfranchised (Caramel 2021).[45]

In an interview about the bilateral deal with Norway, a French journalist asked the new minister about a scandal in the logging sector that cost his

---

41   The Space Observation Agency and the National Resource Inventory were created with resources not earmarked as REDD+. A National Climate Council attached to the Presidency was also established, which is now the focal point for REDD+ related programmes.

42   Interview 4, Remote Sensing Scientist, London, January 30, 2019.

43   Interview 10, Remote Sensing Scientist, Milan, May 14, 2019.

44   When the US White House convened its "Leaders Summit on Climate" in 2021, forty heads of state were invited and Gabon's president was one of them (White House 2021). The only other representative from Central Africa was the president of the Democratic republic of the Congo, a country with eight times more forests and a population forty times larger.

45   Interviews 30 and 37, Environmental Civil Society Organisations 1 and 2, Libreville, August 30, 2019, and September 5, 2019.

predecessor the job (Caramel 2019).[46] The topic, apparently, had not come up in the discussions with the Norwegian partners. In the world of environmental diplomacy, the new minister's scientific credentials could perhaps outshine reports about corruption in the forestry sector, which transnational coalitions of activists try to uncover (e.g., EIA 2019). Illegal logging was, nevertheless, evoked in a technical document detailing the method developed to quantify the carbon losses and gains for the result-based agreement, because it might affect the calculations (CNC 2021, 55). Degradation from selective harvesting—one to five trees per hectare—is hardly detectable from space, and so annual timber production volumes were used in place of remote sensing data to estimate carbon losses and gains in forest concessions. As illegal logging (overexploitation for example) is a known reality, official records could misrepresent actual harvests. According to the document, however, by compiling different information (declared volumes and exports), illegally harvested timber would be "captured" in the numbers (CNC 2021, 56), without having to be first addressed on the ground.

In June 2021, Norway announced the transfer of a first instalment of seventeen million dollars as part of the result-based agreement (CAFI 2021). Only five dollars per tonne were paid for, instead of ten, as the higher price was contingent on methods and audit procedures still being developed by US-based consultants. The report presenting the results indicated that only emissions reductions were counted, not removals, and that a deduction was applied to account for the uncertainty of the estimates (CNC 2020). Carbon losses from deforestation, degradation, and logging in 2016 and 2017 were compared to, and found lower than an average historical level. The agreement aimed to reward improved performance, which may be hard to demonstrate if past performance was good. Given Gabon's low deforestation rates, it had been expected that higher returns could be secured by also rewarding carbon removals (more carbon accumulating in forests remaining forests). But these carbon gains proved disappointing. Carbon storage in 2016 and 2017 was lower than the historical average. The decrease was explained by the early adoption of sustainable management practices in the forestry sector: with less intensively exploited forests, additional carbon uptake from regrowth had slowed down.

Despite being lower than expected, the first result-based payment received by Gabon was celebrated in online media, with photographs of charismatic

---

46   The scandal was referred to as the "Kevazingogate." Shipping containers with, allegedly, illegal kevazingo—a species selling at high prices in China—were seized, and then disappeared. The containers were eventually found but some went missing. The scandal was commented on in national and foreign media (BBC and *Le Monde*).

wildlife (elephants), massive trees, and lush rainforest (Tan 2021).[47] On CNN International, where he was invited to speak, the scientist-minister reasserted with conviction the carbon storage service provided by the country to the planet (CNN International 2021). One may retort that a more effective way for Gabon to mitigate climate change would be to keep its oil in the ground. But even Norway, one of the most affluent societies in the world, does not look ready to do it either (Sutterud and Ulven 2020). In the Autumn 2022, the minister was even confident that Gabon could obtain more per tonne of carbon by selling "sovereign" offsets to companies desperate to compensate for their emissions (Carbon Brief 2022). Whether these plans will come to fruition remains to be seen. Ensuring that deforestation rates stay low, more carbon is stored, national infrastructures are operational to monitor and quantify land-use changes, and international donors maintain a relationship with the government, all depends on decisions and events at home and abroad. The military coup on September 30th 2023 is a sharp reminder that politics can shatter, or at least complicate, lengthily negotiated schemes to valorise the forest's ecological services.

8   Conclusion

To conclude let me go back to the forest near Libreville, where I learned many things about the *Okoume, Azobe, Ozigo, Alep, Okala*, and *Ozouga* trees we passed by—what their wood is used for, the medicinal properties of their bark, roots, sap, and leaves, and their function in rituals. The eco-guide, who spent five years at a research station in the middle of the country assisting ecologists tracking mandrills, kept highlighting the multiple uses of the forest as a "pharmacy" where "everything is organic."[48] Her description of the place echoed the opening chapter of Raponda-Walker's *Les Plantes utiles du Gabon* (Gabon's Useful Plants): "In the vast 'general store' that is the Gabonese forest, even those with the most demanding needs would see them be met!" (Raponda-Walker and Sillans 1995, 30–31).[49] Ethnobotany is, of course, a form of ordering and the Arboretum a place managed for recreational purposes, which large mammals such as forest elephants deserted long ago.[50] Nevertheless, browsing

---

47   The possibility to adjust the performance assessment illustrates the vulnerability to gaming of REDD+ result-based initiatives (Karsenty and Ongolo 2012). See Hook (2020) for a similar discussion in the case of Norway's partnership with Guyana.

48   Arboterum, Libreville, August 25, 2019.

49   Author's translation.

50   On ethnobotany as ordering, see Langwick (2021).

Raponda-Walker's atlas and being guided through the protected area provide an experience of a diverse forest, full of singular trees that contrasts with its quantified existence as stores of carbon.[51]

Walking in the Arboretum did not prompt an extensive conversation about carbon stocks. The metric is too abstract and becomes meaningful on the larger scale when, as shown in this chapter, data from a forest site are connected to more data from distant places, sometimes collected decades ago. Measuring trees is one small, yet essential step in long chains of measurements, calculations, and estimates. These involve allometric equations resulting from dozens of destructive sampling studies and thousands of sacrificed trees. The quantity estimates based on allometric equations can be used to translate land-use change matrices derived from satellite images—themselves the outcome of several billion dollars of investments in earth observation programmes and time-tested remote sensing techniques—into carbon losses and gains. Field measurements and mass estimates may also support the development of algorithms able to process signals recorded by new space-borne instruments. The expectation is that pantropical datasets would be useful to improve the simulation of the vegetation's carbon fluxes in global climate models. Speaking with scientists, I came to understand that every measurement comes with some uncertainty, even at the level of one tree. I then realised that knowing this is already a move towards knowing more about what is happening in those forests, and to our planet, at a time when its habitability is a *real* matter of concern. Making sure that tropical ecosystems continue to clean up our planetary mess is, ultimately, what is at stake in this measurement frenzy, its unequal collaborations *and* interdependencies.

This chapter described funding arrangements, technologies, scientific practices, and political strategies, without which the mass of carbon stored in a forest would not be known, nor would anyone be interested in quantifying it. This led me to foreground big and trivial issues, both essential to the quantification effort—from the possibilities opened by active remote sensing, to tedious procedures for buying food supplies. It allowed me to highlight short-term contingencies (a foreign researcher in forest ecology becoming minister in an autocracy) and long-term phenomena (carbon accumulating in trees over centuries and the diffuse traces of the French colonial project) that must be considered to understand how particular forest lands come to have planetary significance and how this state of affair can suddenly change.

---

51   This echoes Kialo's analysis (2007), which juxtaposes the "economic forest" of French foresters with the forest of Pové people, home to human activities, a diverse fauna and flora, and an invisible domain.

Acknowledgements

I wish to thank those who agreed to spend time with me to share their under-standing of the challenges and implications of quantifying the carbon stored in tropical forests. Any mistakes are mine. I would also like to warmly thank the editors of this volume and the researchers of the African Technoscapes cluster in the Regions2050 project for their close reading of earlier drafts. Their suggestions, and our ongoing conversation, have been extremely help-ful in making the text clearer and sharper. The chapter was discussed at the WiSER seminar, Wits University, in October 2021. This encouraged me to fur-ther tease out the argument and I would like to thank all the participants for their comments. The study was conducted while I was a postdoctoral research fellow at UCL's Institute of Advanced Studies and affiliated to the Department of Geography. The conversations I had with the members of the Institute and the Department directly shaped this research and I am profoundly grateful to them for that.

Bibliography

Angelsen, Arild, Christopher Martius, Veronique De Sy, Amy E. Duchelle, Anne M. Lar-
son, and Pham Thu Thuy. 2018. *Transforming REDD+: Lessons and New Directions*.
Jakarta: Centre for International Forestry Research.

Asase, Alex, Tiwonge I. Mzumara-Gawa, Jesse O. Owino, Andrew T. Peterson, and Erin
Saupe. 2022. "Replacing 'Parachute Science' with 'Global Science' in Ecology and
Conservation Biology." *Conservation Science and Practice* 4, no. 5, e517.

Asiyanbi, Adeniyi P. 2016. "A Political Ecology of REDD+: Property Rights, Militarised
Protectionism, and Carbonised Exclusion in Cross River." *Geoforum* 77, December:
146–56.

Asiyanbi, Adeniyi, and Jens Friis Lund. 2020. "Policy Persistence: REDD+ between Sta-
bilization and Contestation." *Journal of Political Ecology* 27, no. 1: 378–400.

Aubréville, André. 1954. "Premiers Résultats des Plantations d'Okoumé au Gabon." *Bois
et Forêts des Tropiques* 35, May–June: 5–9.

Bernault, Florence, and Tonda, Joseph. 2009. "Le Gabon : Une Dystopie Tropicale." *Politique Africaine* 3, no. 115: 7–26.

Bernard, Philippe, and Christophe Jakubyszyn. 2008. "M. Sarkozy Achève sa Tournée
au Gabon d'Omar Bongo car, 'en Afrique, le Doyen, Cela Compte'." *Le Monde*. Last
modified January 31, 2008. https://www.lemonde.fr/afrique/article/2007/07/28
/m-sarkozy-acheve-sa-tournee-au-gabon-d-omar-bongo-car-en-afrique-le-doyen
-cela-compte_939972_3212.html.

Burton, Mark E. H., John R. Poulsen, Michelle E. Lee, Vincent P. Medjibe, Christopher G. Stewart, Arun Venkataraman, and Lee J. T. White. 2017. "Reducing Carbon Emissions from Forest Conversion for Oil Palm Agriculture in Gabon." *Conservation Letters* 10, no. 3: 297–307.

Caramel, Laurence. 2019. "Lee White: 'Personne n'est Prêt à Payer le Juste Prix pour Sauver les Forêts Tropicales'." *Le Monde*. Last modified September 23, 2019. https://www.lemonde.fr/afrique/article/2019/09/23/lee-white-personne-n-est-pret-a-payer-le-juste-prix-pour-sauver-les-forets-tropicales_6012674_3212.html.

Caramel, Laurence. 2021. "Forêt d'Afrique Centrale : Le Pacte Vert de Lee White." *Le Monde*. Last modified October 29, 2021. https://www.lemonde.fr/afrique/article/2021/10/06/lee-white-un-britannique-au-service-de-la-foret-d-afrique-centrale_6097288_3212.html.

Carbon Brief. 2022. "Cropped, 19 October 2022: Gabon's carbon credits; Living Planet report; Agriculture investigations." Last modified October 19, 2022: https://www.carbonbrief.org/cropped-gabons-carbon-credits-living-planet-report-agriculture-investigations/

CAFI (Central African Forest Initiative).n.d. "Addendum to the Letter of Intent between Gabon and CAFI Signed in 2017 – Results-Based Payment Partnership." Accessed October 28, 2022. https://www.cafi.org/sites/default/files/2022-05/CAFI-%20Gabon%20Addendum%20-%20ENG.pdf.

CAFI (Central African Forest Initiative). 2021. "Gabon Receives First Payment for Reducing CO2 Emissions under Historic CAFI Agreement". Last modified June 22, 2021. https://www.cafi.org/countries/gabon/gabon-receives-first-payment-reducing-co2-emissions-under-historic-cafi-agreement.

Chave, Jérôme, et al. 2005. "Tree Allometry and Improved Estimation of Carbon Stocks and Balance in Tropical Forests." *Oecologia* 145, no. 1: 87–99.

Chave, Jérôme, David Coomes, Steven Jansen, Simon L. Lewis, Nathan G. Swenson, and Amy E. Zanne. 2009. "Towards a Worldwide Wood Economics Spectrum." *Ecology Letters* 12, no. 4: 351–66.

Chave, Jérôme, et al. 2014. "Improved Allometric Models to Estimate the Aboveground Biomass of Tropical Trees." *Global Change Biology* 20, no. 10: 3177–90.

Cheyns, Emmanuelle, Laura Silva-Castañeda, and Pierre-Marie Aubert. 2020. "Missing the Forest for the Data? Conflicting Valuations of the Forest and Cultivable Lands." *Land Use Policy* 96, 103591.

CNN International. 2021. "Gabon Gets Paid $17 Million for Lowering Carbon Emissions." *Facebook CNN Africa*. https://www.facebook.com/CNNAfrica/videos/gabon-gets-paid-17-million-for-lowering-carbon-emissions/507149200489020/.

CNC (Conseil National Climat). 2020. "Gabon National Results Report: Result-Based Payments under the Central African Forests Initiative – Gabon Partnership." https://www.cafi.org/sites/default/files/2021-06/Gabon%20National%20Results%20Report_submitted_18Dec2020.pdf.

CNC (Conseil National Climat). 2021. "Gabon's Proposed National REDD+ Forest Reference Level." https://redd.unfccc.int/files/gabon_frl_submitted_feb2021.pdf.

Crane, Johanna T. 2013. *Scrambling for Africa: AIDS, Expertise, and the Rise of American Global Health Science*. Ithaca: Cornell University Press.

Disney, Mathias. 2019. "Terrestrial LiDAR: A Three-Dimensional Revolution in How We Look at Trees." *New Phytologist* 222, no. 4: 1736–41.

Dougueli, Georges. 2019. "Gabon : Dix Choses à Savoir sur Lee White, Ministre des Forêts, de la Mer et de l'Environnement." *Jeune Afrique*. Last modified June 18, 2019. https://www.jeuneafrique.com/mag/789009/politique/gabon-dix-choses-a-savoir -sur-lee-white-ministre-des-forets-de-la-mer-et-de-lenvironnement/.

Edwards, Paul N. 2010. *A Vast Machine: Computer Models, Climate Data, and the Politics of Global Warming*. Cambridge, MA: MIT Press.

Ehrenstein, Véra. 2018. "Carbon Sink Geopolitics." *Economy and Society* 47, no. 1: 162–86.

EIA (Environmental Investigation Agency). 2019. "Toxic Trade: Forest Crime in Gabon and the Republic of Congo and Contamination of the US Market." Last modified March 25, 2019. https://eia-global.org/reports/20190325-toxic-trade.

Fairhead, James, and Melissa Leach. 1995. "False Forest History, Complicit Social Analysis: Rethinking Some West African Environmental Narratives." *World Development* 23, no. 6: 1023–35.

Fatoyinbo, Temilol et al. 2021. "The NASA AfriSAR Campaign: Airborne SAR and Lidar Measurements of Tropical Forest Structure and Biomass in Support of Current and Future Space Missions." *Remote Sensing of Environment* 264, 112533.

Fayolle, Adeline et al. 2018. "A Regional Allometry for the Congo Basin Forests Based on the Largest Ever Destructive Sampling." *Forest Ecology and Management* 430, December: 228–40.

Fearnside, Philip M. "Science and Carbon Sinks in Brazil." *Climatic Change* 97, no. 3: 373–78.

Gabrys, Jennifer. 2020. "Smart Forests and Data Practices: From the Internet of Trees to Planetary Governance." *Big Data & Society* 7, no. 1: 1–10.

GEDI (Global Ecosystem Dynamics Investigation). 2019. "GEDI Status – Feb 1." Last modified February 1, 2019. https://gedi.umd.edu/gedi-status-feb-1/.

Geissler, Paul W. 2013. "Public Secrets in Public Health: Knowing not to Know while Making Scientific Knowledge." *American Ethnologist* 40, no.1: 13–34.

Goldstein, Jenny E. 2022. "More Data, More Problems? Incompatible Uncertainty in Indonesia's Climate Change Mitigation Projects." *Geoforum* 132, June: 195–204.

Gupta Aarti, Eva Lövbrand, Esther Turnhout, and Vijge Marjanneke J. 2012. "In Pursuit of Carbon Accountability: The Politics of REDD+ Measuring, Reporting and Verification Systems." *Current Opinion in Environmental Sustainability* 4, no. 6: 726–31.

Hecht, Gabrielle. 2018. "Interscalar Vehicles for an African Anthropocene: On Waste, Temporality, and Violence." *Cultural Anthropology* 33, no. 1: 109–41.

Hermansen, Erlend A.T., and Sjur Kasa. 2014. "Climate Policy Constraints and NGO Entrepreneurship: The Story of Norway's Leadership in REDD+ Financing." CGD Working Paper 389, Washington: Center for Global Development.

Herold, Martin et al. 2019. "The Role and Need for Space-based Forest Biomass-related Measurements in Environmental Management and Policy". *Surveys in Geophysics* 40, no. 4: 757–78.

Hook, Andrew. 2020. "Following REDD+: Elite Agendas, Political Temporalities, and the Politics of Environmental Policy Failure in Guyana." *Environment and Planning E: Nature and Space* 3, no. 4: 999–1029.

Hountondji, Paulin. 1990. "Scientific Dependence in Africa Today." *Research in African Literatures* 21, no. 3: 5–15.

Hubau, Wannes, et al. 2020. "Asynchronous Carbon Sink Saturation in African and Amazonian Tropical Forests." *Nature* 579, 80–7.

Karsenty, Alain, and Symphorien Ongolo. 2012. "Can 'Fragile States' Decide to Reduce their Deforestation? The Inappropriate Use of the Theory of Incentives with Respect to the REDD Mechanism." *Forest Policy and Economics* 18: 38–45.

Kialo, Paulin. 2007. *Anthropologie de la Forêt: Populations Pové et Exploitants Forestiers Français au Gabon*. Paris: L'Harmattan.

Kwa, Chunglin. 2005. "Local Ecologies and Global Science: Discourses and Strategies of the International Geosphere-Biosphere Programme." *Social Studies of Science* 35, no. 6: 923–50.

Lahsen, Myanna. 2009. "A Science-Policy Interface in the Global South: The Politics of Carbon Sinks and Science in Brazil." *Climatic Change* 97, no. 3–4: 339–72.

Lang, Chris. 2019. "Lee White NYDF 2019." YouTube video, 9:18. https://www.youtube.com/watch?v=AKcmwHl3n_I.

Langwick, Stacey. 2021. "Properties of (Dis) Possession: Therapeutic Plants, Intellectual Property, and Questions of Justice in Tanzania." *Osiris* 36, no. 1: 284–305.

Latour, Bruno. 1999. *Pandora's Hope: Essays on the Reality of Science Studies*. Cambridge, MA: Harvard University Press.

M'Bokolo, Elikia. 1981. *Noirs et Blancs en Afrique Équatoriale : Les Sociétés Côtières et la Pénétration Française, (vers 1820–1874)*. Paris: *Éditions* de l'EHESS.

Mallard, Alexandre. 1998. "Compare, Standardize and Settle Agreement. On Some Usual Metrological Problems." *Social Studies of Science* 28, no. 4: 571–601.

Mavhunga, Clapperton C. 2017. "Introduction." In *What do Science, Technology, and Innovation mean from Africa?*, edited by Clapperton C. Mavhunga, 1–27. Cambridge, MA: MIT Press.

Mbembe, Achille. 2007. "L'Afrique de Nicolas Sarkozy." *Mouvements* 52, no. 4: 65–73.

Mbembe, Achille. 2020. *Brutalisme*. Paris: La Découverte.

Mouissi, Mays. 2018. "Impacts des Activités du Groupe Olam sur l'Économie de la République Gabonaise entre 2010–2017." Mays Mouissi Consulting. Accessed October 28, 2022. https://www.mays-mouissi.com/wp-content/uploads/2018/05

/ETUDE-Impacts-des-activite%CC%81s-du-Groupe-Olam-sur-l%E2%80%99e
%CC%81conomie-de-la-Re%CC%81publique-gabonaise-entre-2010-2017-Mays
-Mouissi-Consulting.pdf.

Mouity, Patrice M. 2012. "L'Émergence Réformiste, ou comment Rompre et Perpétuer l'Etat Bongoïste." In *Le Gabon à l'Épreuve de la Politique de l'Émergence: Diagnostic et Prognostic,* edited by Patrice M. Mouity and Kévin-Ferdinand Ndjimba, 29–53. Paris: Publibook.

NASA (National Aeronautics and Space Administration). 2016. "NASA, Partner Space Agencies Measure Forests in Gabon." Last modified February 25, 2016. YouTube video, 3:25. https://www.youtube.com/watch?v=n1hxI53rq5Q.

Ngomanda, Alfred et al. 2014. "Site-Specific *Versus* Pantropical Allometric Equations: Which Option to Estimate the Biomass of a Moist Central African Forest?" *Forest Ecology and Management* 312: 1–9.

O'Connell, Joseph. 1993. "Metrology: The Creation of Universality by the Circulation of Particulars." *Social Studies of Science* 23, no. 1: 129–73.

Okeke, Iruka N. 2016. "African Biomedical Scientists and the Promises of 'Big Science'." *Canadian Journal of African Studies/Revue Canadienne des Etudes Africaines* 50, no. 3: 455–78.

Ongolo, Symphorien, and Alain Karsenty. 2015. "The Politics of Forestland Use in a Cunning Government: Lessons for Contemporary Forest Governance Reforms." *International Forestry Review* 17, no. 2: 195–209.

Osseo-Asare, Abena Dove. 2014. *Bitter Roots: The Search for Healing Plants in Africa.* Chicago: University of Chicago Press.

Pan, Yude et al. 2011. "A Large and Persistent Carbon Sink in the World's Forests." *Science* 333, no. 6045: 988–93.

Poulsen, John R. et al. 2020. "Old Growth Afrotropical Forests Critical for Maintaining Forest Carbon." *Global Ecology and Biogeography* 29, no. 10: 1785–98.

Pourtier, Roland. 1989. *Le Gabon, Tome* 2. Paris: L'Harmattan.

Raponda-Walker André, and Roger Sillans. (1961) 1995. *Les Plantes Utiles du Gabon.* Libreville: Fondation Raponda-Walker.

Rottenburg, Richard. 2009. "Social and Public Experiments and New Figurations of Science and Politics in Postcolonial Africa." *Postcolonial Studies* 12, no. 4: 423–40.

Saatchi, Sassan S. et al. 2011. "Benchmark Map of Forest Carbon Stocks in Tropical Regions Across Three Continents." *Proceedings of the National Academy of Sciences* 108, no. 24: 9899–904.

Sallenave, Pierre. 1955. *Propriétés Physiques et Mécaniques des Bois Tropicaux de l'Union française.* Nogent-sur-Marne: Centre Technique Forestier Tropical.

Soir 3 Journal. 2007. "Nicolas Sarkozy au Gabon." Last modified July 27, 2007. Soir 3 Journal video, 2:10. https://www.ina.fr/ina-eclaire-actu/video/3402166001004/nicolas-sarkozy -au-gabon.

Sutterud, Tone and Elisabeth Ulven. 2020. "Norway Plans to Drill for Oil in Untouched Arctic Areas." *The Guardian*. Last modified August 26, 2020. https://www.theguardian.com/environment/2020/aug/26/norway-plans-to-drill-for-oil-in-untouched-arctic-areas-svalbard.

Tan, Jim. 2021. "Gabon Becomes First African Country to Get Paid for Protecting its Forests." *Mongabay*. Last modified July 20, 2021. https://news.mongabay.com/2021/07/gabon-becomes-first-african-country-to-get-paid-for-protecting-its-forests/.

Tousignant, Noémi. 2018. *Edges of Exposure: Toxicology and the Problem of Capacity in Postcolonial Senegal*. Durham, NC: Duke University Press.

Turnhout, Esther, Aarti Gupta, Janice Weatherley-Singh, Marjanneke J. Vijge, Jessica de Koning, Ingrid J. Visseren-Hamakers, Martin Herold, and Marcus Lederer. 2017. "Envisioning REDD+ in a Post-Paris Era: Between Evolving Expectations and Current Practice." *WIRE s Clim Change* 8, no.1: e425.

Verran, Helen. 2002. "A Postcolonial Moment in Science Studies: Alternative Firing Regimes of Environmental Scientists and Aboriginal Landowners." *Social Studies of Science* 32, no. 5–6: 729–762.

Verran, Helen. 2010. "Number as an Inventive Frontier in Knowing and Working Australia's Water Resources." *Anthropological Theory* 10, no.1–2: 171–78.

Viard-Crétat, Aurore. 2015. *La Déforestation Évitée. Socio-Anthropologie d'un Nouvel « Or Vert ». Entre Lutte contre le Changement Climatique et Aide au Développement, du Laboratoire Guyanais à l'Expertise Forestière au Cameroun*. Doctoral dissertation, Paris: Ecole des hautes études en sciences sociales.

Walford, Antonia. 2012. "Data Moves: Taking Amazonian Climate Science Seriously." *The Cambridge Journal of Anthropology* 30, no. 2: 101–17.

Walford, Antonia. 2017. "Raw Data: Making Relations Matter." *Social Analysis* 61, no. 2: 65–80.

Walker, Anthony P. et al. 2021. "Integrating the Evidence for a Terrestrial Carbon Sink Caused by Increasing Atmospheric CO_2." *New Phytologist* 229, no. 5: 2413–45.

Walters, Gretchen and Melis Ece. 2017. "Getting Ready for REDD+: Recognition and Donor-country Project Development Dynamics in Central Africa." *Conservation and Society* 15, no. 4: 451–64.

Walters, Gretchen et al. 2016. "Peri-Urban Conservation in the Mondah Forest of Libreville, Gabon: Red List Assessments of Endemic Plant Species, and Avoiding Protected Area Downsizing." *Oryx* 50, no. 3: 419–30.

The White House. 2021. "President Biden Invites 40 World Leaders to Leaders Summit on Climate." Last modified March 26, 2021. https://www.whitehouse.gov/briefing-room/statements-releases/2021/03/26/president-biden-invites-40-world-leaders-to-leaders-summit-on-climate/.

CHAPTER 3

Privacy, Privation, and Person: Data, Debt, and Infrastructured Personhood

Emma Park and Kevin P. Donovan

1 Introduction

In early September 2020, a self-described "diaspora" Kenyan in Malawi, Brian Munyao Longwe, wrote to a mailing list for Kenyans working at the intersection of computer technology and public policy. "So—mimi niko na issue (I have an issue)," he began, "My Safaricom number 0715964281 has apparently been repossessed and sold to someone else." As a result, when Longwe had tried to send money to his daughter in Nairobi using Safaricom's mobile money app, M-Pesa, he was shocked to see his phone number assigned to a "Beatrice Chelangat." First turning to the "Kenyans in Malawi" WhatsApp group, he learned that his "line … [was] gone" due to the policy of Safaricom (Kenya's largest mobile network operator) of reassigning numbers after a customer has gone six months without purchasing airtime. This was not the first time Longwe had gone so long without topping up his account—after all, he doesn't make calls or text on Safaricom while abroad. But it "always 'wakes up' when I load airtime," he said, even if it is "six, seven, eight months without a topup [sic] … What gives?" he asked the WhatsApp group.[1]

For Longwe, what he referred to as Safaricom's "reappropriation" of his line was not merely an inconvenience. It was also a threat to his identity, his social relations, and perhaps even his claim to Kenyanness. "I am distressed," he wrote, "because my M-Pesa, my e-Citizen, NTSA, bank accounts, and many other digital assets/identity related items are linked to this number which I have had for the past fifteen plus years. My digital identity (and that of many others in similar predicament) is at risk." This sequence of linked assets and identities constitute core infrastructures of citizenship and exchange for contemporary Kenyans. M-Pesa, the ubiquitous mobile money service, was evidently crucial to his relationship with his daughter; however, the phone number also anchored his relationship with the Kenyan state through its e-Citizen

---

1 Brian Munyao Longwe, "Safaricom Repossessing Numbers (What the heck!)," *Kictanet,* available at https://lists.kictanet.or.ke/pipermail/kictanet/2020-September/036170.html

initiative—through which Kenyans access government services—and the National Transport & Safety Authority drivers' licenses. Even his banks, he explained, knew him through his phone number.

Other members of the mailing list quickly concurred, noting with frustration how they found themselves in similar predicaments. Years ago, it was not such a problem, observed Simiyu, but now it was much more serious. As more and more of life was routed through mobile phones, it was incumbent on the telephone company to better notify people of the impending loss of their account: "I mean, we get birthday texts, it is not any more difficult to implement." Perhaps it was a job for the regulator, some suggested. Others on the group chat thought it contravened the 2019 Kenya Personal Data Protection Act—shifting Longwe's number to another customer unfairly made inaccessible the data Safaricom held in custody for its customer.

Still others chimed in noting the careful maintenance work they undertake to avoid losing their line. "The secret," said a Kenyan based in the United States for twenty years, is just loading enough airtime and doing one small transaction like buying airtime once a month! Without this expenditure, you risked your number being "rudely assigned to someone else." Maria chimed in, writing "I feel for you Ndugu Longwe ... That one, fight it out *hata kama ni KORTINI* (even if it is in COURT!)." After all, she concluded, with a play on Safaricom's advertisements, "My SAFCOM MY *MAISHA* (LIFE) MY IDENTITY!"[2]

This chapter proceeds from technopolitical controversies and anxieties like that of Brian Munyao Longwe to inquire into the status of personhood in contemporary Kenya. Today, firms such as Safaricom play an increasingly large role in structuring not only society and the economy but also, as we argue here, shaping personhood. While his difficulties are perhaps mundane in the digital realm—who has not been locked out of an account, after all?—situating them within the anthropological and historical literature on personhood offers new ways to understand not only the predicaments of digital infrastructures but also the contours of social relations in Kenya and beyond. Our goal is not merely to analyse the meaning of mobile technology in a particular setting or culture—though we foreground vernacular idioms and significations; it is also to insist that digital technologies across the continent and beyond oblige analysts to reassess presumptions about obligation, individuality, and personhood that indeed shape much of the writing on the topic.

While literature concerned with infrastructures has focused on large-scale sociotechnical systems to explore processes of nation-building (Mukerji

---

2 A prominent Safaricom advertising campaign asserts that "Maisha ni Digital" (Swahili for Life is Digital).

2009), racialized forms of dispossession (Karuka 2019), and labour (Mains 2013; Anand 2017), less attention has been directed towards the question of how infrastructures are implicated in the process of subjectification (cf. von Schnitzler 2016)—that is the constitution of persons. In the present case we argue that as the traces of people and their networks are routed through infrastructural platforms one can see the heightened importance of what we define below as *infrastructured personhood*.

In working towards an understanding of how digital infrastructures are intermediating people and the social relations that constitute them, we primarily focus on one arena where mobile telephony has particularly dramatised the tensions and contradictions of personhood, namely the use of mobile phones for lending to low-income Kenyans. Digital loans have exploded in popularity since 2015 in Kenya. By monitoring calls, texts, mobile money transactions, and location, dozens of companies are creating credit scores for users who then receive loans via Kenya's mobile money network, M-Pesa (Donovan and Park 2019). In our fieldwork, users' concerns about privacy and publicity subtended their more overt anxieties about debt and privation. Yet, these worries are not unique to digital debt. The past decade has played host to a series of largely unresolved predicaments that hinge on the control of information, the distribution of authority, and the dynamics of personhood and identity. Here we draw on the ethnographic and historical record, as well as insights from STS, to shed light on how new digital infrastructures are both operationalising and reconfiguring distributions of personhood and the forms of attachment and detachment—from kin, from technologies—that this new dispensation portends. In doing so, we emphasise how these novelties implicate long-standing tensions between different sorts of being in Kenya and beyond.

2 Distributed Personhood and Digital Data

A core concern in what follows is how digital infrastructures animate an enduring contrast between modes of personhood. Anthropological analyses of personhood have largely been a comparative project of contrasting the (putatively Western) "individual" with its cultural others, whether the Melanesian "dividual" (Strathern 1988) or a sub-Saharan "relational" variant (Comaroff and Comaroff 2001). These discussions usually begin with Marcel Mauss's (1985 [1938]) lecture on "the category of the person" which pointed to the variety of ways in which selves and types are construed; related discussions find their home in Foucauldian scholarship on the making of social types (e.g., Hacking 2006). African philosophers, too, emphasise that "in the African view" of personhood, "it is the community which defines the person as person, not some isolated static quality of rationality, will, or memory." It is therefore more processual, "not given simply because one is born" (Menkiti 1984, 171–72). We broadly share the concern with understanding how ideas and practices of personhood are assembled, yet, we agree with recent writing that departs from an anthropological tradition that maps styles of personhood onto distinct cultural regions and instead, recasts the "dividual"/"individual" contrast as the basis for differences and conflicts *within* particular social worlds (Zoanni 2018, 63, emphasis added; see also Bialecki and Daswani 2015). Likewise, we think of personhood not as something you *have* so much as something you *achieve*, a sort of process that is "realized over time and thus vulnerable to blockages and disruptions in the course of particular lives" (Zoanni 2018, 64–65).

In other words, we are concerned with competing practices and processes of assembling personhood within Kenya. We see personhood as the result of particular ensembles of language, law, ethics, and infrastructures. We track the contradictions and tensions within Kenyan society between "the individual"—an autonomous being detached from others—and what we call *the infrastructured person*—one constituted through sociotechnical attachments. While our latter term shares much with discussions of "relational personhood" (Comaroff and Comaroff 2001), we prefer the term *infrastructured* for the way it points to the materiality of attachments through which personhood is composed.

These forms of personhood are both normative ideals and practical achievements. They exist (to varying degrees latent or explicit) throughout everyday life, legal doctrine, and technopolitical projects in Kenya (as well as elsewhere). The tensions between personhood formatted as individuality and personhood formatted as relationality is, we think, key to understanding recent anxieties and scandals surrounding digital technology. As Brian Munyao Longwe's predicament suggests, digital infrastructures do not simply mediate questions of access—though they certainly do this as well—nor have they displaced extant forms of sociality. Crucial is how they intermediate and assemble persons and the social relations that constitute everyday life.

3   Persons, Selves, and Actor-Networks

It might be worth noting a certain similarity between this ethnographic theory and the most prominent social theory within STS: within actor-network theory, too, there is a focus on the composition of entities through their relations. Bruno Latour (2005), Michel Callon (1984), and Annemarie Mol (2003), among others, emphasised that stabilised entities are the result, not the premise, of action. While *personhood,* or the *self,* have not been ANT's terms of analysis, the manner in which actors are constructed is at the core of the theory. Callon and Latour's (1981) classic reworking of Hobbes's notion of Leviathan, for instance, emphasises that the capacity of a person is dependent upon their enlisting a sociotechnical network of others, including non-human things. Likewise, Latour (1983) argues that Pasteur could not have developed a vaccine to protect against anthrax without enrolling allies in the form of farmers, the bacillus, laboratory instruments, and norms of practice. It is these sociotechnical alliances that give Leviathan, Pasteur, or any actor their capacity and standing. And while ANT's effort has been to foreground the distribution of agency, this does not mean we cannot speak of Pasteur or the sovereign as acting, once we carefully trace the deployment of their network of relations and account for the inevitable translation of action at a distance. Those that we call individual actors, in other words, are those who can marshal reliable relations.

The affinities between actor-network theory and anthropological accounts of personhood were the subject of an early analysis and important critique of Latour by Marilyn Strathern (1996). Strathern's own theorisation of Melanasian "dividuals" is among the most influential explications of personhood outside the modern legal fiction of an autonomous individual. She rightly noted the shared characteristics of ANT and her school of thought, but argued that actor-networks are not endlessly extensive. At a certain point, they are "cut," no longer enrolling allies. How is this so? One crucial way is through the law, especially property law, through which states allocate access to assemblages to some and foreclose their availability to others. Property regimes, in other words, *enclose* actor-networks by limiting their enrolment of (or by) others.

Yet, the law is not the only mode of structuring the extension and delimitation of sociotechnical ties. Everyday action involves an inordinate range of deferral, dismissal, avoidance, and escape through which people limit their relationality. We discuss some of these below: window curtains and locked gates, hedgerows and hidden herds, missed calls and ignored SMSs, blocked phone numbers, and claims to have no money to lend. Once you start to look for the ways of being related and ways of being alone, they are everywhere. What matters is not so much the dance of enclosure and exposure, attachment and detachment—that abstraction is general enough. Rather, this awareness should direct ethnographic attention towards the particular practices, techniques, idioms, and implications of these dynamics.

4   Contemporary Tensions of Infrastructured Personhood in Kenya

To do so, we turn in this section to some contemporary examples of this interplay. Particularly noteworthy are those that involve mobile telephony and mobile money, both of whose ubiquity today should not distract from their rarity a mere two decades ago. These infrastructures do not only yoke people together in ways commonplace to networked technologies; they also facilitate various practices of detachment, through which morally normative social relations have been placed under strain. For instance, while the mobile money service M-Pesa is famous for facilitating domestic remittances, many rural residents complain that their urban kin now visit less often, choosing to send money rather than to deliver it themselves. What it means to be a good son—the relations and exchanges, movement and gestures that compose it—are shifting, with digital infrastructure variously impeding and affording these practices.

If the relations among users are most evident, mobile phones and the services built upon them have also inaugurated a range of other attachments, especially between users and corporations. Today, Kenyans are daily engaged as customers and users with companies like Safaricom, which is able to translate its considerable user base into the most profitable earnings of any firm in the region. This dispensation also encourages the distribution of persons across the firm's sociotechnical networks. One's identity, "reputational collateral," and standing are enmeshed in Safaricom's infrastructure (Breckenridge 2019, 95). This happens in ways that today are completely ordinary, such as having one's phone number painted on their business, rather than one's name or address. These infrastructural traces are read by Kenyans with a fine interpretive repertoire. They know which prefixes are for which mobile network operators and use such knowledge to save money on intra-network calling. They know that a mobile number starting with 0722 is not only a Safaricom line, but also that it was the first batch of mobile numbers in Kenya, marking the owner of that number as not only a prestigious early adopter but also someone rich enough in the early 2000s to buy a SIM card—and to never lose it through inactivity. Giving your number, in other words, isn't a merely quantitative exercise, but for those with a 0722 line, it can be an act of proudly marking one's status in the world.

And as the opening example to this chapter makes clear, the loss of one's phone line and number can threaten one's standing in the world, cutting one off from the set of relations and networks through which a person exists. Such a loss is not uncommon, including through the inordinately common practice of SIM swap fraud, through which third parties take over someone's phone number through subterfuge (such as claiming to have lost their phone and needing a replacement SIM card). One survey suggested a quarter of Kenya's forty-three million mobile phone users had been victims of this practice (CapitalFM 2018). In Kenya, this is particularly violating because a SIM card is also a wallet for mobile money. Here, the concern is that the extension of one's identity—the practical activities through which personhood is constructed—across networked technologies renders people vulnerable to a perforation of their personhood, not merely their privacy.

The "infrastructural attachments"—through which people and their life-worlds become tethered to sociotechnical networks (Park 2017)—that concern us most in what follows is the collection of intimate, granular data about mobile phone users. In the past five years, the data collected from the ordinary operation of these telecommunications infrastructures has grown in importance, becoming a commodity to be sold and used to refine other corporate products. As a result, to use Haggerty and Ericson's (2000) term, there exist an ever-growing number of "data doubles" for the abstraction of human identity and behavior into consolidated archives. In databases operated by mobile operators and third-party apps, Kenyans' *clicks*, *likes*, movement, and transactions are monitored, collected, and classified.

For some, this reflects an affront to individual privacy (e.g., Nyabola 2018), leading to calls for a Data Protection Act (inaugurated in 2019) that governs "personal data," which the act defines as "information relating to an identified or identifiable natural person" (Government of Kenya 2019). But what strikes us is the widespread belief, shared by corporate lawyers and many digital rights advocates alike, that digital data is *personal* rather than social. As some scholars have pointed out (e.g., Davies 2015), the networked nature of these products means digital data is more often than not about relations, not individuals. Who you call or text, your mobile money transaction history, and even the posts you *like* or comment upon implicate others. While the Kenya Data Protection Act and associated legal forms speak of a "data subject" as a "natural person" the subject of digital data is, to a considerable extent, what we have called an infrastructured person—relational, or distributed, and not an autonomous individual. Social media corporations, of course, understand this, speaking of the value of the social graph in Facebook's case, and seeking to amass as many users' relational data as possible, rather than focusing on profits from individuals. The law, however, has as its "object and purpose" to "regulate the processing of personal data" and to "protect the privacy of individuals" (Government of Kenya 2019). Just as private property is the law's normative frame of reference for a world of commodities, so too is the autonomous individual subject both the object of legal framing and the agent of their own actions. This conceit, though backed by the law, fails to capture the distribution of personhood in Kenya and beyond (Viljoen 2020).

Indeed, the economy of digital data is not one where the commodity implicates autonomous individuals, at least not uniquely. This was not the case for

Brian Munyao Longwe, whose individual identification markers were important not for his autonomous self but, rather, for how they mediated his relations
with his daughter, his banks, and the Kenyan state. Nor is it the case in the
subject of our extended discussion—digital lending—where data collected
and analyzed is not so much about a borrower as individual person or market participant, but rather about borrowers as socially situated, relational persons whose digital activities and borrowing implicates a person constituted by
relations across a sociotechnical world. The tensions between individual and
relational persons is, perhaps, best revealed in the ethnographic and historical
mode, and it is to these that we now turn.

5   Digital Data and Debt

In the summer of 2019, we were in Nairobi to conduct interviews about popular
involvement in the stock market, but what our friends and interlocutors really
wanted to talk about was a different type of finance: digital lending. Time and
again, somewhat stilted conversations on stock shareholding and initial public
offerings lit up at the mention of M-Shwari, Fuliza, Tala, Branch or any number of dozens of phone-based lending services. In recent years, these FinTech
(or financial technology) apps have proliferated. This became clear during
our very first conversation in a cafe sharing tea and *mandazi* with a group of
University of Nairobi students.[3] Talk turned to digital lending and what many
called a "crisis" in consumer overindebtedness (fieldnotes, June 2019). On this
and subsequent occasions conversation moved easily from the strains this crisis was putting on interpersonal relations, to a discussion of the ways FinTech
firms come to digitally know would-be borrowers, to questions regarding the
role that government should play in protecting citizens.

Mainly youth on the cusp of social adulthood are being hurt by these
apps, one of our mid-career interlocutors opined, explaining that he came to
understand this when a cousin asked for help repaying two long-overdue digital loans.[4] Between two FinTech firms, he owed KSh30,000. This was largely
due to a gambling habit that led this well-to-do guy—a medical student in his
fourth year of study—to spend the school fees allocated for his children. His
habit had "eaten the school fees," his cousin lamented. The result was "very
embarrassing," both for him and for family members to whom he was forced
to appeal for help. Note that while this medical student had achieved some

---

3   *Mandazi* are small fried pastries that people often eat with sweet, milky tea.
4   Fieldnotes, Nairobi, June 2019.

of the markers of social adulthood—namely children and a job that theoretically allowed him to provision for his family—his reliance on the larger family unit to backstop his debt threatened to undo this hard-won achievement. Put simply, for some, indebtedness is not merely a source of embarrassment but is existential, threatening to throw aspiring social adults back in developmental time, rendering them, once again, social juniors.

Some we spoke to during fieldwork thought recklessness—exemplified by an urge to consume rather than save and produce—explained over-indebtedness. But not all of the fault was to be placed on defaulters, many people we spoke to argued.[5] Part of the issue was the infrastructure of lending, which capitalises on the ubiquity of mobile phones. The apps are not passive in aggregating users; they send routine nudges, advising people that to get a loan, you "just need to dial this number." One interlocutor recounted how FinTech firms and banks alike send text messages advertising loans as "easy [cheap] credit," rather than as their opposite, expensive "debt." As for the aspiring doctor, he had been able to access such relatively large loans because of the modalities by which FinTech algorithms had come to digitally know him as a potential borrower, namely by monitoring his digital behavior. Through these dubious means, he argued, the algorithms can "tell the kind of person I am ... It's a very mad thing," he concluded, "[t]he government should protect its citizens."

Versions of this narrative were reported to us myriad times over the course of the summer. Lenders were referred to as dogs, devils, and fools, evidence of a growing critique of the industry. Debt was called a source of embarrassment and shame for the indebted. Borrowers were referred to as "slaves," tricked by firms who "give you money gently, and then ... come for your neck."[6] Such laments indicate the ways people's presents and futures are subordinated to the rhythms of borrowing and repayment dictated by the logic of FinTech apps rather than the more elastic and labile forms of credit offered through interpersonal borrowing—of which we will have more to say below.

For some, the problem was one of privacy. Writing in one of the leading dailies, Franklin Sunday and Macharia Kamau explained that millions of Kenyans are "knowingly or unknowingly surrendering their privacy as well as their contact list of those they communicate with in exchange for exorbitant mobile loans" (Sunday & Kamau 2019). They continued, "other conditions set by online lenders that are oblivious to mobile loan users include giving up a pound of flesh for instant cash, a lifetime of SMS notifications, full surrender of their personal data to third parties and a waiver of their right to dignity." If their tone

5   Fieldnotes, Nairobi, June 2019.
6   Fieldnotes, Nairobi, June 2019.

was a bit hectoring, Sunday and Kamau were not alone in their worry about the privacy implications of these apps. We heard versions of this lament time and again, such as one man, a Nairobi borrower, who spoke of the firms "tracking digital footprints and financial transactions to get a picture of you."[7]

The most infamous case was that of OKash (though they are not alone), which endeavored to compel repayment from borrowers by purposefully shaming them. Remotely accessing the phone contact list of a borrower—who "agreed" to such intrusion when they took the loan—OKash's collection agents would look for influential entries (e.g., "Mom" or "Boss") and call them to report the borrower's lack of repayment. When this started happening, Kenyans were appalled, taking to social media and the newspapers to condemn the practice. Others found different workarounds: one man explained to us that people began putting the collection agents in their phones under entries like "Devil" or "Do Not Pick Up."[8]

A young Nairobi man, Valentine, explained to us that he thought the whole strategy was not merely offensive—it was likely useless. "If someone called me about my friend's debts," he speculated, "there is no way I would get involved. There's only trouble that way".[9] For Valentine, the appropriate ethical stance— the safe one—was one of detachment. Tom Neumark has observed something similar among poor women in eastern Nairobi: giving and receiving are dangerous, opening one to excessive obligation as well as being perceived to be a burden. Neumark's (2017, 749) interlocutors sought to "disentangle themselves from others while cultivating personal ... economic self-sufficiency." For our contact Valentine, entanglement in the financial lives of others was likely to end in trouble, a situation exacerbated if one allowed themselves to be deputised by an aggressive debt collector. Indeed, Valentine insisted he took a rather detached approach to the whole digital borrowing craze: "I cannot afford to be in debt," he laughed, explaining that he has so far avoided it. Yet, for many others, debt is a fact of life—not least because it's impossible to afford to *not* be in debt. This is true in a narrow financial sense, as these lending apps have been enrolled by Kenyans struggling to make rent, pay school fees, and support their households (Donovan and Park 2019).

---

7   Interview with borrower, Nairobi, June 2019.

8   Fieldnotes, Nairobi, June 2019. Remarkably, the way he learned this was similar to the transgressive act OKash was doing. Many people in Kenya use an app called TrueCaller, he explained, which mines all users' contact lists to guess the name of an unknown caller. So, a TrueCaller user who gets a call from an unknown number will be presented with a suggestion based on the entries in others' phone contact lists: "Maybe Marcel Mauss." But note the similarity: both OKash and TrueCaller were mining phone contacts, while only the former were considered to offend moral sensibilities.

9   Ibid.

It is also true in a more expansive, social sense, and in the next section we broaden the frame to illustrate the manifold ways this is true.

6   Attachment, Detachment, and Personhood in Kenyan History

In this section, we step back from the present-day examples to examine the historical and anthropological record to demonstrate the long-standing tension between interpersonal obligations and individual autonomy. The discussions below may seem a far stretch from the topic of this chapter, which regards the rise of digital debt and its relation to data in contemporary Kenya. Yet our goal is to place contemporary transformations of personhood—too often reduced to technological and legal novelties—within an enduring cultural context. If such a move is less common in STS scholarship, it is the bread and butter of anthropologists and historians who seek to understand how ethical norms and moral commitments shape exchanges, both commercial and non-commercial. Anthropologist Parker Shipton (2007) calls these "fiduciary cultures" to point to how wealth, credit, and debt are bound up in cultural values. Others speak of "moral economies" (Rogan 2019). Our goal in depicting at some length these social dynamics in various Kenyan settings is to eventually return to the more overtly infrastructural dynamics of contemporary digital debt, and the moral uncertainties of this particular market (cf. Callon and Muniesa 2005).

In Kenya, as elsewhere, interpersonal dynamics are, from the outset, shaped by dynamics of attachment and detachment with various others. Historically, mutual support and various forms of insurance have been embedded in relations of kinship rather than being delegated to third parties, as they may be in social democratic settings. The centrality of mutuality is the product of long-standing norms of maintaining and deepening social attachments. Consider building a family. The birth of an infant marks a moment wherein the connections between the past (the ancestors), the present (living kin), and the future (the unborn), are articulated, marking the extension of relationality rather than its abrogation. Raising children is a costly affair in emotional, financial, and psychological terms. Remittances from the city to the countryside are one modality through which these incrementally accreted debts, debts which can never be fully repaid, are mediated (Gugler 2002; Ross and Weisner 1977).

The case is even clearer in the instance of marriage. In patrilineal societies, once a male child has gone through initiation—a symbolic rite that marks the formal transition from boy to man—the father and father's brothers are expected to aid the initiate in securing a wife by assembling bridewealth payments. This, too, entails both dynamics of credit and debt where kinship networks are bound together in an ongoing fashion rather than transactional exchanges premised on a logic of finality. Herds are broken apart as livestock move from the family of the groom to the family of the bride. These payments often take place over the course of many years, sometimes extending to the deaths of parties to the couple. As Jean and John Comaroff note for Southern Africa, this is a protracted and incremental process wherein the question is not whether "two people were married, but how much" (Comaroff and Comaroff 2001, 271). Increasingly in Kenya, bridewealth payments involve the mobilization of small contributions of myriad kin and friends (Park 2017; Kusimba 2018). This not only binds together kin groups, but sutures men and women to their aging parents, who expect their children to subsidise their livelihoods through cash remittances in old age (Mintz-Roth and Heyer 2016). Such transfers sustain the elderly in the absence of a pension.

These repertoires of imbrication, however, are matched by (and sometimes compete with) dynamics of detachment and individuation. For instance, among Kikuyu people historically, self-actualisation was reflected in structures of authority. Upon being circumcised, a man would seek to separate himself from his father's *shamba,* or garden, cutting a section of forest and labouring to produce a tillable plot. This piece of cultured land was the first step towards establishing a household. Once a given homestead grew too large, frontier groups reached outward, taking over neighbouring ridges (Lonsdale 1992). This was a profoundly future-oriented theory of value. It was only by virtue of pursuing land, marriage, and children that a man would be remembered beyond his death; a man's legacy being sustained through his children and their mothers, and the generations that succeeded them. Similarly, among the Wakamba, pride of place was given to men who distinguished themselves from others by demonstrating the particularities of their skills cultivated through mobility. In the nineteenth century, large-game hunting, esoteric knowledge of the wider world, and access to objects of prestige were all metrics of self-actualisation premised on an ethics of distinction and individuation (Osborne 2014).

This dynamic of attachment and detachment continued over the course of a life-cycle, wherein it was (and often is) through establishing relations of both (inter)dependence and of distinction—of disembedding oneself as one aggregates followers—that one achieves socially sanctioned personhood. For Gikuyu speakers, this is captured linguistically by the term *wiathi,* or self-mastery, which is achieved by successfully parlaying one's wealth in things—namely livestock and land—into wealth in people, while separating oneself from the homestead of one's father (Lonsdale 1992; Peterson 2008). If adulthood involves the diminishing of certain dependencies on one's parents, it likewise involves becoming a patron for others, most notably one's own children but wider networks of kin (fictive and otherwise) as well. In other words, leadership is premised on a politics of redistribution, a debt owed as the appropriate return for loyalty. Such obligations could fuel resentments and resistance.

Any reckoning with these problematics must attend to both the relational attachments and the individuating detachments. These social obligations often fuel resentment and resistance. The overbearing burdens of kin have motivated Kenyans to seek more autonomous existence. Moving to the city has been one way Kenyans historically escaped the strictures of these kinship structures, finding in the relative anonymity of Nairobi or Mombasa or Kisumu a space for self-making outside the moral expectations of elders and others (White 1990). But other modes existed too. Mission schools offered young Kenyans spaces for personal advancement beyond the rules of family (Peterson 2004), and the wages of agricultural labour opened new avenues for consumption that could accelerate one's achievement of social standing (Ocobock 2017). In their ethnographic study with Dholuo speakers, Paul Wenzel Geissler, and Ruth Jane Prince emphasise the importance of "practices of relatedness and ethical reflections about relations" (210, 8) yet also note ways of embodying more "individual 'self-discipline'" (210, 83) through identifying as a "Saved" Christian. Here, religious practices serve to disentangle believers from relations in which they might otherwise be imbricated.

Our point is not to analytically define individual autonomy so much as to point to the experience of escape from social obligation. Even in these new settings, historians document, Kenyans made new attachments (White 1990; Cooper 1987), and their detachments from what came before could carry significant risks. It could mark you not merely as socially inferior, but in some instances could relegate you to a position outside the social collective entirely.

7 Matters of Borrowing and Lending

For many in Kenya, livestock and land are both stores of wealth and future-oriented asset classes geared toward securing wealth in people: cattle because they are translated into bridewealth payments braiding together kinship groups, and land because of the value placed on establishing a household with a garden. The symbolic weight placed on these investments helps make them secure. But it is also their materiality that shapes their status: unlike cash, cattle cannot easily be physically rendered partible in the event that a family member is in need. Moreover, for many in the region, cattle were not historically subject to easy commodification. Instead, they were intimates, or what Julie

Livingston (2019, 42) has referred to as "inter-speciated familiars." Land, for its part, was often protected from parceling out and commodification due to prohibitions that viewed selling the land of the ancestors as severing ties with the dead, with dangerous implications for the living and the unborn (Shipton and Goheen 1992; cf. Li 2014).

That these forms of investment were considered useful because they were "classificatory wholes" in their material forms (Guyer 2016, 157), has not, however, precluded the possibility of lending and borrowing (Shipton 2007). Kin might be allowed to tend a part of a herd on their own land, reaping the benefits of the milk generated (Shipton 2007, Chapter 4). These loans often extend over long periods of time and build up connections between people across space. The spatial component also protects the security of the investment by mitigating against the risk of an entire herd being decimated by the routine spread of zoonotic diseases (Spear and Waller 1993). In Kenya's highlands, cattle transgressed ethnic boundaries, leading to Kikuyu farmers saving their returns in livestock, what Lonsdale (2012, 39) called the "reserve currency of reputation and power." Maasai, the "bankers to the highland market" (Lonsdale 2012, 27), drew upon historical exchange—including of kin—in moments of need, when they provided work to Kikuyu in return for agricultural products (Ambler 1988).

These styles of resource allocation composed certain types of personhood. Among the Maasai in the Kenya-Tanzania borderlands, it was historically women who had claim to milk in return for tending the herd (Hodgson 2001, 224); here, as elsewhere, careful labour justified claims-making. As for land, indigent neighbors or friends might be invited to till a plot of land—these people are called *ahoi* in Gikuyu and *jodak* in Dholuo—from which they reap the vegetables they grow and are given the authorisation to construct structures and assemble life (Cohen and Odhiambo 1989; Lonsdale 1992). Sometimes these temporary arrangements extended over many generations, and may translate into permanent residence; this being particularly likely if there is a marriage that binds together the family of the tenant and the landholder (Lonsdale 1992). Alternatively, those struggling to get by will be invited to aid during periods of harvest, taking a quantity of the bounty as their payment (Cohen and Odhiambo 1989). In the cities, indigent family members or neighbors living in the countryside are often brought in to work as house-help, in exchange for which they secure a roof over their heads and transport to the city.

Not all are invited to partake in these networks and assessments of character determines one's capacity to be a borrower. Put differently, what, how, and on what terms you borrow is contingent on shifting interpersonal relationships not only between the borrower and the lender, but between their families in the past and the potential relations between them in the future. This is, in part, because the temporal horizons that characterise these tethers of credit and debt are not dictated by the foreshortened timeline of financial institutions, which expect a monetary return on a loan, but are governed by the value accreted through more tightly weaving the one to the other. These relations of credit and debt, put simply, are socially productive. This is not to naively posit that these dynamics are uncomplicated—they are structured by often quite firm hierarchies of gerontocratic and patriarchal power and authority—it is simply to note that they are productive of value that is not foremost monetary in character. Moreover, relations of indebtedness turn on the knowledge that one is always liable to move from the position of creditor to that of debtor, a dialectical dynamic that ensures the never seamless reproduction of the normative social order. The socially successful person is always defined in relation to relations with others.

Shipton (2007), based on his work among Luo communities in western Kenya, has referred to this dynamic of relationality as "entrustment" to describe the mutuality of credit and debt relations—at times serial and at others reciprocal—that bind kin across time and space. For example, unhappy kin (living or dead) are routinely cited as the cause of financial struggles (Shipton 1989). Forgotten rites or unintended slights can have serious, intergenerational consequences. In this fiduciary culture, relations of debt and credit are rarely imagined as binding two individuals to each other as individuals. Nor are these imagined as relations which are severed once a debt has been paid (as contract law would have it) but, instead, work as a social glue by virtue of their capacity to bring nested scales of interconnection into greater imbrication. Yet, as Shipton insists, these intimacies and entanglements are liable to spark conflict and schism.

These expectations are normative. And people who do not conform to this notion of relational personhood—evidenced by, for example, not investing in their rural communities after moving to cities or by driving around the countryside with your car windows rolled up, an act read as a rejection of the community that raised you (Park 2017)—quickly tip into accusations of anti-sociality. Such accusations generate a range of responses from generalised suspicion to accusations regarding witchcraft and sorcery. One might say that the only thing more dangerous than overbearing social obligation is individual autarky.

There is, in other words, a balance between what Zebulon Dingley (Unpublished MS) calls "enclosure" and "exposure." Writing about suspicion and witchcraft on the South Coast of Kenya, Dingley emphasises the risks of relatedness as well as the importance of being embedded within networks of kin and neighbors. The risks come, in part, due to a sense that one cannot know the intentions of others, and that both intimates and strangers can present a threat in a context of a limited amount of resources and status. This obliges everyday interpretive work and practical actions to both open oneself to the relations through which proper social standing are achieved, as well as to take steps to enclose oneself from malevolent intentions, actions, and forces.

In our experience living and working in other parts of the country, the dynamics Dingley describes are not unique to the South Coast, even if the forms of exposure and enclosure are particular. Others have also noted these dynamics. On the other side of the country, Günter Wagner's (1949, 178) study of North Kavirondo documented an elaborate range of ritual and medicinal measures for protecting against the malevolent actions of others, living and dead. Resources, too, were enclosed from too much reciprocity, including through prohibitions on letting out land to non-heirs (Wagner 1956, 78). Among Luo communities, hedges and thorn bushes historically separated homes, enforcing boundaries that were crossed by kin and neighbors at the invitation of the host. In more recent decades, wealthier residents have stood apart from others through more durable, less permeable concrete walls and metal gates, chancing opprobrium from the community. Even the "precariously open" and permeable body must be managed lest it become too vulnerable (Geissler and Prince 2010, 166–67).

Similarly, among the Gikuyu of central Kenya, the pinnacle of successful personhood—namely, the homestead—was historically guarded from the peering eyes, jealous feelings, and malintentions of others through the construction of intricately woven hedges (Peterson 2004, 11). This was captured linguistically, the verb—*gita* meaning to "grow thick, close as a hedge," a process which was linked to the future, the verb also meaning to flourish or prosper.[10] Today in Nairobi, while during the day apartment doors are open as neighboring children move through the compound, at night curtains are tightly drawn to guard against curious neighbors and unwelcome others. Newly purchased items that are the subject of display and pride during daylight hours become sources of jealousy once the sun has fallen. Mobile phones are used to be connected just as often as they are employed to avoid connections; missing calls, texts, and requests for remittances are all tactics of detachment that people pursue (cf. Archambault 2016). In other words, notions of "entrustment" and "relationality" are riven with uncertainty, suspicion, and anxiety.

Establishing oneself as a person who exists through his or her relations with other persons thus requires cultivating an openness, such that relations of entrustment can be forged, while fortifying oneself against the unknowable intentions and threatening actions of others; which itself requires occluding aspects of the self that might be the source of moral and social censure.

---

10   Thanks to Derek Peterson for sharing this insight.

Attachment always depends on correctly calibrating its other: detachment. While the historical archive is clear that tensions between these notions have a long history—animated by shifting legal orders, property regimes, and labour practices—they have been renewed in the past fifteen years as digital infrastructures expanded in Kenya.

8   Navigating between Financial and Social Debts in Digital Kenya

Having detoured through the historical ethnography of these norms and practices within Kenya, we can now better understand the social contours of money and finance today. Kenyan ideas regarding personhood not only allow us to see with greater clarity the business models of FinTech firms, but also gives us purchase on the framework of critique brought to bear by users. Despite promises otherwise, Kenyan debtors see loans offered by FinTech firms less as ushering in new regimes of individuation. Instead they are more concerned with how firms—and the platforms upon which they depend—are intermediating social life by generating what we refer to as infrastructured persons. This analytic points to both the centrality of specific sociotechnical entanglements that underwrite FinTech, as well as highlighting that these processes are best understood not as a wholesale rupture but, instead, build on extant relational networks of expectations of credit and debt. It is precisely this mediation that permits the appropriation of value produced socially (Bear et al. 2016).

Kenyans find their place in fiduciary cultures an ambivalent one.[11] Sibel Kusimba (2021, 109–127) writes of efforts to avoid being asked or having to give

---

11   The ambivalence of attachment also holds for *chamas*, the local term for rotating credit and savings groups, where groups of peers pool their contributions. The topic of frequent anthropological and developmental inquiry, *chamas* exemplify the ambiguities of collective credit and debt. The terms of borrowing can be flexible, often responding to the needs of a given group member, a repertoire built over the course of many hours of intimate knowing. The support offered in such contexts is often not simply financial, moreover, with group members offering moral as well as material support. Without access to bank loans, *chamas* provide a manner of accumulating usefully large lump sums, with no fees or intimidating paperwork. As late as 2018, informal financial groups, such as chamas, were used by 41 percent of the Kenyan population, while only 35 percent of the population held traditional bank accounts. Yet, group entanglement offers its own discipline, with the cost of default including social opprobrium and malice. In her study of similar practices in South Africa, Deborah James (2015: 121) notes that savings clubs are riven by a 'lack of trust,' with the possibility of someone absconding with the wealth always hovering on the edges. In other instances, jealousy between members can cause disruption, even dissolution of the group.

financial assistance. Turning off a phone, missing calls, or feigning ignorance are ways to avoid obligations to others. Yet, she notes the risks of escaping too much, including fostering resentment and losing resources you may need down the line. Such burdens are not evenly distributed, either, with women often tasked with redistributing through social networks what resources exist. Here, for better or worse, gendered expectations around provisioning and responsibility can place women at the center of struggles over getting by. The appeal of digital debt emerges in the context of this fiduciary culture. In place of interpersonal negotiations, FinTech offers the seemingly impersonal decisions of the algorithm. Instead of the vulnerability, shame, or indignity of asking a family or friend for money, "these apps," writes one journalist, "have enabled those in need to take the loans quietly, saving them the embarrassment of borrowing from friends ... or begging savings groups members to offer them loans" (Xinhua 2019).

Digital debt, on this reading, offers Kenyans something appealing: the possibility of financial credit without *social* debt. By attaching themselves to the algorithms of financial technologies, Kenyans might gain a measure of individual autonomy, bypassing the complications of interpersonal relations and relational personhood. It is a promise of fashioning selves without the encumbrances of others—a detaching from some others, though, entails different, commercial obligations (cf. Kusimba 2021, 135). This is, we have argued, an older story in Kenya, one iconically captured in the wage earning young person, whose employment and urban life made them relatively autonomous from the dictates and resources of rural elders (Ocobock 2017; White 1990).

It is also a familiar story under contemporary capitalism—including its financial varieties—where people are hailed as *individual* workers, consumers, and borrowers. In Wendy Brown's (2016) influential account of neoliberalism in general, but particularly in Europe, this is a story of collective life being undone through the formation of individual accumulators, entrepreneurs of the self. In our conversations with Kenyan colleagues, friends, and interlocutors, however, this (neo)liberal fantasy has not been borne out. People spoke not of a process of individuation, giving rise to the detached person, but of novel distributions of the self as new modalities of relationality are enacted by digital platforms.

This was the case in at least two ways. First, our interlocutors acknowledge that FinTech inaugurates new forms of attachment to corporations and their infrastructures. As a result, they do their best to attach themselves in beneficial ways, such as topping up their balance or enrolling friends, practices that improve their standing with the apps and their algorithms. If credit scoring algorithms are often opaque, it hardly stops efforts to comport oneself in ways befitting its judgment—efforts to secure what Keith Breckenridge has usefully referred to as "reputational collateral" (2019, 95). For their part, the lending firms recognize the social implications of debt and try to harness peer pressure and shame to compel borrowers to repay.

Secondly, Kenyans explain the new digital debt as activating and relying upon social relations. Time and again, we heard stories of overly indebted borrowers turning to cousins and friends for help—cashing in, so to speak, on entrustment to cover their financial obligations. Many borrowers who find themselves unable to repay must turn to friends or family for assistance. Lelei, for instance, told us how she defaulted on a digital debt that she originally borrowed for her cousin who had, themselves, already gotten into debt trouble with their phone. Lelei confessed that she expected her cousin to never repay, and she doubted she could come up with the cash herself, but she nevertheless borrowed. Why, we asked? She paused, trying to find the way to convey the depth of her relation with this cousin: they have "taught me many skills. We've done life together," and this is a debt greater than any monetary value.[12] If relational personhood is, in part, an achievement built through active collaboration with others, then Lelei's standing reflects the contributions of her cousin. This much she recognized in her decision to sacrifice another element of her assembled status—namely, her credit score—by borrowing on behalf of her kin. So, the promise of individualised borrowing often turns out to rely on a pre-existing entrustment among borrowers. Channeling Marxist feminist scholarship, we might even say the usurious creditors are subsidised by social relations and the interpersonal (and largely non-fiscal) debts people draw upon (cf. Dalla Costa 1975; Federici 2020).

To these two ways that relational personhood is reflected in digital debt, we would add a third. While the digital apps seemingly offered a version of "asocial assistance" (Ferguson 2015), the conditions of possibility for borrowing were premised precisely on the algorithmic analysis of borrowers' social relations. Put simply, the firm's profitability depends on the reserves of relational personhood's obligations. FinTech firms invest considerably in knowing these social dynamics—from ethnography to digital scraping—in order to modulate the credit offered (Donovan and Park 2022(b) and 2022(c)). In other cases, they actively position themselves as a relation to which one owes repayment by dint of goodwill rather than legal obligation: as we were told by an industry insider, people repay Safaricom loans, in part, because of the goodwill the firm cultivates through philanthropy.

---

12   Interview with borrower, Nairobi, June 2019.

To put it simply, the digital lending industry is not at all divorced from everyday social relations. While from the perspective of a borrower, such loans forego the need to supplicate to intimates, the lending algorithms and their designers are themselves keenly concerned with the relationality of Kenyans. The data that informs the credit scoring of these apps is not data about autonomous, self-directed individuals. It is data about their relations and interactions.

Consider the elementary form of digital traces used to make many FinTech credit scores: Call Detail Record (CDR). The CDR is generated by SIM cards connected to mobile phone networks; they provide a unique record of phone activities, including the caller ID, the recipient ID, the cell tower from which the call was placed and from which it was received, the time of the call and its duration. Similar transactional information is generated through the use of mobile money like M-Pesa: sender, recipient, time, location, and amount. While these data are personal they also are social for they are relational. They almost always implicate others: calls and remittances are, after all, transactions between people. Credit scoring algorithms monitor who you interact with and who they interact with, qualifying your standing on the basis of your relations' standing.

Further, activities that may not seem interpersonal classify people socially, because "men and women tend to use their phones differently, as do different age groups. Frequently making and receiving calls with contacts outside of one's immediate community is correlated with higher socio-economic class," for instance (UN Global Pulse 2013, 6). Even your location is meaningful only because of the value of those co-located with you: a SIM card that spends nights in a poor neighborhood is surrounded by SIM cards whose transactional data reveals the commercial limits of their users.

The individual, then, is a thoroughly social product for digital lenders. It is not merely that individuals are of a social category—urban youth, say, or secondary school female with a smart phone—as marketing professionals have long recognised. Rather, it is that in the monitoring of reputation, FinTech construes borrowers as relational persons, whose digital data is not so much the result of an individual person's behavior as it is the trace of an infrastructured self, composed socially and distributed across digital networks and filtered through algorithms. Here, individual privacy offers little consolation.

Bibliography

Ambler, Charles H. 1988. *Kenyan Communities in the Age of Imperialism: The Central Region in the Late Nineteenth Century.* New Haven: Yale University Press.

Archambault, Julie Soleil. 2016. *Mobile Secrets Youth, Intimacy, and the Politics of Pretense in Mozambique.* Chicago: University of Chicago Press.

Bear, Laura, Karen Ho, Anna Lowenhaupt Tsing, and Sylvia Yanagisako. 2015. "Gens: A Feminist Manifesto for the Study of Capitalism." Theorizing the Contemporary, *Fieldsights*. Last modified March 30, 2015. https://culanth.org/fieldsights/gens -a-feminist-manifesto-for-the-study-of-capitalism.

Bialecki, Jon and Girish Daswani. 2015. "Introduction: What is an Individual? The View from Christianity." *HAU: Journal of Ethnographic Theory* 5, no. 1: 271–94.

Breckenridge, Keith. 2019. "The Failure of the 'Single Source of Truth about Kenyans': The NDRS, Collateral Mysteries and the Safaricom Monopoly". *African Studies* 78, no. 1: 91–111.

Brown, Wendy. 2016. "Sacrificial Citizenship: Neoliberalism, Human Capital, and Austerity Politics." *Constellations* 23, no. 1: 3–14.

Callon, Michel. 1984. "Some Elements of a Sociology of Translation: Domestication of the Scallops and the Fishermen of St Brieuc Bay." *The Sociological Review* 32, no. 1: 196–233.

Callon, Michel and Bruno Latour. 1981. "Unscrewing the Big Leviathan: How Actors Macro-Structure Reality and How Sociologists Help Them to Do So." In *Advances in Social Theory and Methodology: Toward an Integration of Micro-and Macro- Sociologies*, edited by Karin Knorr Cetina and Aaron Victor Cicourel, 277–303. Boston: Routledge and Kegan Paul.

Callon, Michel and Fabian Muniesa. 2005. "Economic Markets as Calculative Collective Devices." *Organization Studies* 26, no. 8: 1229–50.

CapitalFM. 2018. "Over 10m Mobile Users in Kenya hit by SIM Fraud." Last modified October 24, 2018. https://www.capitalfm.co.ke/business/2018/10/over-10m-mobile -users-in-kenya-hit-by-sim-fraud/.

Carrithers, Michael, Steven Collins and Steven Lukes, eds. 1985. *The Category of the Person: Anthropology, Philosophy, History*. Cambridge: Cambridge University Press.

Cohen, David William and E. S. Atieno Odhiambo. 1989. *Siaya: The Historical Anthropology of an African Landscape*. Nairobi: Heinemann.

Comaroff, Jean and John L. Comaroff. 2001. "On Personhood: An Anthropological Perspective from Africa." *Social Identities: Journal for the Study of Race, Nation and Culture* 7, no. 2: 267–83.

Cooper, Frederick. 1987. *On the African Waterfront: Urban Disorder and the Transformation of Work in Colonial Mombasa*. New Haven: Yale University Press.

Dalla Costa, Mariarosa and Selma James. 1975. *The Power of Women and the Subversion of the Community*. Bristol: Falling Wall Press Ltd.

Davies, Will. 2015. "The Return of Social Government: From 'Socialist Calculation' to 'Social Analytics.'" *European Journal of Social Theory* 18, no. 4: 431–50.

Dingley, Zebulon. Unplubished MS. "Enclosure and Exposure in South Coast, Kenya: Body, House, Settlement."

Donovan, Kevin P. and Emma Park. 2019. "Perpetual Debt in the Silicon Savannah." *Boston Review*. Last modified September 20, 2019. https://bostonreview.net/class

-inequality-global-justice/kevin-p-donovan-emma-park-perpetual-debt-silicon
-savannah.

Donovan, Kevin P. and Emma Park. 2022(a). "Algorithmic Intimacy: The Data Economy
of Predatory Inclusion in Kenya." *Social Anthropology/Anthropologie Sociale* 30, no.
2: 120–39.

Donovan, Kevin P. and Emma Park. 2022(b). "Knowledge/Seizure: Data, Debt & Rent
in Kenya," Antipode 54, no. 4

Donovan, Kevin P. and Emma Park 2022(c). "Algorithmic Intimacy: The Data Economy
of Predatory Inclusion in Kenya," Special Issue on "Curious Utopias: Large and Small
Blueprints for Human Society," Social Anthropology 30, no. 2

Federici, Silvia. 2020. *Revolution at Point Zero: Housework, Reproduction, and Feminist
Struggle*. Oakland: PM Press.

Ferguson, James. 2015. *Give a Man a Fish: Reflections on the New Politics of Distribution*.
Durham, NC: Duke University Press.

Geissler, Paul Wenzel and Ruth Jane Prince. 2010. *The Land Is Dying: Contingency, Creativity and Conflict in Western Kenya*. New York and Oxford: Berghahn Books.

Government of Kenya. 2019. The Data Protection Act (No. 24 of 2019). http://kenyalaw
.org/kl/fileadmin/pdfdownloads/Acts/2019/TheDataProtectionAct__No24of2019
.pdf.

Gugler, Josef. 2002. "The Son of the Hawk Does Not Remain Abroad: The Urban – Rural
Connection in Africa." *African Studies Review* 45, no. 1: 21–41.

Guyer, Jane I. 2016. *Legacies, Logics, Logistics: Essays in the Anthropology of the Platform
Economy*. Chicago: University of Chicago Press.

Hacking, Ian. 2006. "Making Up People." *London Review of Books* 16, no. 28.

Haggerty, Kevin D. and Richard V. Ericson. 2000. "The Surveillant Assemblage." *British
Journal of Sociology* 51, no. 4: 605–22.

Hodgson, Dorothy. 2001. *Once Intrepid Warriors: Gender, Ethnicity and the Cultural Politics of Maasai Development*. Bloomington: Indiana University Press.

James, Deborah. 2015. *Money from Nothing: Indebtedness and Aspiration in South Africa*.
Stanford: Stanford University Press.

Karuka, Manu. 2019. *Empire's Tracks: Indigenous Nations, Chinese Workers, and the
Transcontinental Railroad*. Oakland: University of California Press.

Kusimba, Sibel. 2018. "Money, Mobile Money and Rituals in Western Kenya: The Contingency Fund and the Thirteenth Cow." *African Studies Review* 61, no. 2: 1–25.

Kusimba, Sibel. 2021. *Reimagining Money: Kenya in the Digital Finance Revolution*. Culture and Economic Life. Stanford: Stanford University Press.

Latour, Bruno. 1983. "Give Me a Laboratory and I Will Raise the World." In *Science
Observed: Perspectives on the Social Study of Science*, edited by K. Knorr-Cetina and
M.J Mulkay, 141–170. London and Beverly Hills: Sage Publications.

Latour, Bruno. 2005. *Reassembling the Social: An Introduction to Actor-Network-Theory*.
Oxford and New York: Oxford University Press.

Livingston, Julie. 2019. *Self-Devouring Growth: A Planetary Parable as Told from Southern Africa*. Durham, NC: Duke University Press.

Lonsdale, John. 1992. "The Moral Economy of Mau Mau: Wealth, Poverty and Civic Virtue in Kikuyu Political Thought." In *Unhappy Valley: Conflict in Kenya and Africa. Book Two: Violence and Ethnicity,* edited by Bruce Berman and John Lonsdale, 315–468. London: James Currey.

Lonsdale, John. 2010. "Ethnic Patriotism and Markets in African History." *JICA Research Institute,* no. 20.

Mains, Daniel. 2013. *Hope Is Cut: Youth, Unemployment, and the Future in Urban Ethiopia*. Philadelphia: Temple University Press.

Mauss, Marcel. (1938) 1985. "A Category of the Human Mind: the Notion of Person, the Notion of Self." In *The Category of the Person: Anthropology, Philosophy, History,* edited by Michael Carrithers, Steven Collins and Steven Lukes. Cambridge: Cambridge University Press.

Menkiti, Ifeanyi A. 1984. "Person and Community in African Traditional Thought." In *African Philosophy: An Introduction,* edited by Richard A. Wright, 171–182. University Press of America.

Mintz-Roth, Misha and Amrik Heyer. 2016. " 'Sharing Secrets': Gendered Landscapes of Trust and Intimacy in Kenya's Digital Financial Marketplace." In *Trusting and its Tribulations: Interdisciplinary Engagements with Intimacy, Sociality and Trust,* edited by Vigdis Broch-Due and Margit Ystanes, 131–147. Brooklyn: Berghahn Books.

Mol, Annemarie. 2002. *The Body Multiple: Ontology in Medical Practice*. Science and Cultural Theory. Durham, NC: Duke University Press.

Mukerji, Chandra. 2009. *Impossible Engineering: Technology and Territoriality on the Canal Du Midi*. Princeton Studies in Cultural Sociology. Princeton, NJ: Princeton University Press.

Neumark, Tom. 2017. "'A Good Neighbour is Not one That Gives': Detachment, Ethics, and the Relational Self in Kenya." *Journal of the Royal Anthropological Institute* 23, no. 4: 748–64.

Nyabola, Nanjala. 2018. *Digital Democracy, Analogue Politics: How the Internet Era is Transforming Kenya*. London: Zed Books.

Ocobock, Paul. 2017. *An Uncertain Age: The Politics of Manhood in Kenya*. Athens: Ohio University Press.

Osborne, Myles. 2014. *Ethnicity and Empire in Kenya: Loyalty and Martial Race Among the Kamba, c. 1800 to the Present*. Cambridge: Cambridge University Press.

Park, Emma. 2017. "Infrastructural Attachments: Mobile Entrepreneurs and the Tensions of 'Home' in Colonial and Postcolonial Kenya." Dissertation. Ann Arbor: University of Michigan.

Peterson, Derek R. 2004. *Creative Writing: Translation, Bookkeeping, and the Work of Imagination in Colonial Kenya*. Portsmouth, NH: Heinemann.

Peterson, Derek R. 2008. "The Intellectual Lives of Mau Mau Detainees". *The Journal of African History* 49, no. 1: 73–91.

Rogan, Tim. 2017. *The Moral Economists: R.H. Tawney, Karl Polanyi, E.P. Thompson, and the Critique of Capitalism*. Princeton, NJ and Oxford: Princeton University Press.

Ross, Marc Howard, and Thomas S. Weisner. 1977. "The Rural-Urban Migrant Network in Kenya: Some General Implications." *American Ethnologist* 4, no. 2: 359–75.

Von Schnitzler, Antina. 2016. *Democracy's Infrastructure: Techno-Politics and Protest after Apartheid*. Princeton, NJ: Princeton University Press.

Shipton, Parker and Mitzi Goheen. 1992. "Introduction. Understanding African Land-Holding: Power, Wealth, and Meaning." *Africa: Journal of the International African Institute* 62, no. 3: 307–25.

Shipton, Parker. 1989. *Bitter Money: Cultural Economy and Some African Meanings of Forbidden Commodities*. Washington, DC: American Anthropological Association.

Shipton, Parker. 2007. *The Nature of Entrustment: Intimacy, Exchange, and the Sacred in Africa*. New Haven: Yale University Press.

Spear, Thomas T., Richard Waller, eds. 1993. *Being Maasai: Ethnicity & Identity in East Africa*. Eastern African Studies. London and Dar es Salaam and Nairobi and Athens: James Currey and Mkuki na Nyota and EAEP and Ohio University Press.

Strathern, Marilyn. 1996. "Cutting the Network." *Journal of the Royal Anthropological Institute* 2, no. 3: 517–35.

Strathern, Marilyn. 1988. *The Gender of the Gift: Problems with Women and Problems with Society in Melanesia*. Oakland: University of California Press.

Sunday, Franklin and Macharia Kamau. 2019. "How Kenyans Give Up Privacy for Costly Mobile Loans." https://www.standardmedia.co.ke/financial-standard/article /2001310291/how-kenyans-give-up-privacy-for-costly-mobile-loans.

Taylor, Keeanga-Yamahtta. 2019. *Race for Profit: How Banks and the Real Estate Industry Undermined Black Homeownership*. Chapel Hill: The University of North Carolina Press.

UN Global Pulse. 2013. Big Data for Development: A Primer.

Viljoen, Salome. 2020. "Democratic Data: A Relational Theory For Data Governance." SSRN Scholarly Paper. Rochester, NY: Social Science Research Network. Accessed September 15, 2022.

Wagner, Gunter. 1949. *The Bantu of North Kavirondo*. Oxford: Oxford University Press.

Wagner, Gunter. 1956. *The Bantu of North Kavirondo: Economic Life*. International African Institute.

White, Luise. 1990. *The Comforts of Home: Prostitution in Colonial Nairobi*. Chicago: University of Chicago Press.

Xinhua. 2019. "Digital Lenders Offer Broke Kenyans Relief After Festive Season". Last modified January 7, 2019. http://www.xinhuanet.com/english/2019-01/07/c _137726471.htm.

Zoanni, Tyler. 2018. "The Possibilities of Failure: Personhood and Cognitive Disability in Urban Uganda." *The Cambridge Journal of Anthropology* 36, no. 1: 61–79.

CHAPTER 4

'Graduates-as-a-Service': Running a Data Factory in Northern Uganda

René Umlauf

1   Introduction

During the months from December to April, a period of the year that is also known as the dry season, temperatures in the northern Ugandan city of Gulu often rise above forty degrees Celsius. On the campus of Gulu University, individual seminars are then moved outdoors. The cool shade of dense mango and acacia trees is often the only option for lecturers and students to escape the stifling heat of the university buildings covered with iron roofs. This kind of improvised teaching practice formed the contrasting background to the colourfully decorated and air-conditioned containers of Samasource. As I had learned shortly before, Samasource is a San Francisco-based company that specialises in producing accurate data by combining machine assisted annotation (MAA) with human validation for Machine Learning (ML) and Artificial Intelligence (AI). After passing the guarded doors, it felt as though I was entering a spaceship that was about to take off because of the constant whirring of the air-conditioner amplified by the noise of approximately 150 computer ventilators. The environment was intensified by the tightly closing windows of the containers, which not only muffled noise from outside, but also ensured that cool air would not escape the two hundred square metre room. "We couldn't work here without A/C!," explained William, the manager, "With windows open, tonnes of dust would enter and quickly destroy the PCs. But also, excess heat from the desktops would make it unbearable. And of course, we all know that cooled bodies perform better ..."[1]

The manager's quick reflections on both the climatic peculiarities, and the insight that cooled bodies would achieve higher productivity than uncooled ones, again underlined the extreme contrast not only to the rest of the campus architecture, but also to the predominantly rural character of the rest of

---

1   Interview #2, Manager Samasource, Gulu, September 19, 2018.

northern Uganda. As a non-profit organisation, Samasource uses special soft-ware applications to break down very large datasets into small components, which are subsequently processed and labelled by a large number of work-ers. For such annotation, or micro-work—such as image tagging or semantic segmentation—workers only need a basic knowledge of classical literacy and numerical skills as well as a basic command of digital literacy.

Based on this ethnographic experience, in the following contribution I will elaborate on some conditions that I believe are central for the successful oper-ation and general functioning of this comparatively young form of data factory work. I will argue that the (infra)structured connection to the World Wide Web and the access it facilitates to digital labour is based on a number of different but interacting *disconnections*. This chapter delineates the emerging social-technical tensions and asks which disconnections, points of access, and circu-lations of digital labour these tensions are based upon. Before offering further ethnographic details, the paper situates the conceptual discussions around "micro-work" by critically linking it to a broader debate centred around the concept of the digital divide (see Selwyn and Facer 2010, Selwyn 2004).

In the first empirical part of this paper, I will analyse the organisational effects and causes of the disconnections of Samasource operations from local realities, with emphasis on the infrastructural and material aspects. In the sec-ond part, examples will be presented, indicating how an extremely selective allocation of and access to digital resources represent a central operational condition for Samasource. In this context, software-supported control and standardisation regimes form an essential prerequisite for decoupling individ-ual activities from relying on local knowledge and working practices. In the third and final part, a contrast is made between wage labour, which is based on strictly standardised measuring of labour, and the specific requirements of local labour cultures that are characteristic of low resource settings.

The aim of this paper is to make visible the specific conditions on which novel forms of globally circulating labour rely when they travel to places like the University of Gulu. To do so, it will be necessary to highlight organisational, infrastructural, social, and knowledge-based decouplings that contribute, to varying degrees, to how these new forms of digital labour overlap, and at times conflict with local work cultures. However, while digital evolutions in the urban centres of African countries, for example Nairobi, Accra, or Lagos have received growing academic attention, literature on the networking of periph-eral rural regions is rather sparse; this paper is therefore an attempt to broaden the field of research in this regard.

2 Micro-work, Digital Divide and the Resource Character of the Internet

The notion "humans-as-a-service"[2] is used to describe all those activities in digital economies for which there are either no technical or automated solutions (yet) available, or for which it is not deemed economically worthwhile to ever develop a technical solution (Irani 2013, 11; Bergvall-Kareborn and Howcroft 2014). Content managers, for instance, who clean up platforms from violent images and hate speech represent a sore spot, an embarrassment to Silicon Valley solutionism[3] that is not able to offer a technical fix for emerging ethical and political problems. Content work in the Philippines and California are often outsourced to other companies and is not granted the same payment nor the luxurious workplace conditions that Facebook or Google engineers enjoy. Humans-as-a-service demarcates the distinction between what Silicon Valley considers real work (building platforms) versus simple content work (Daub 2020). In a similar vein, big car companies like BMW or General Motors keep suspiciously silent when entering into contracts with big data factories in China and India for the production of large chunks of training data for their self-driving cars. But being silent about this heavy reliance on cheap human labour to make automation and AI work, is paralleled by a fear of diminishing and disenchanting public imaginaries of the technology's own intelligence (Gray and Suri 2019).

On the one hand, the spectrum of activities and tasks subsumed under the notion of humans-as-a-service is very broad and becomes increasingly difficult to compare. A few minutes of text transcription on Amazon's "mechanical turk"[4] is very different from the psychological stress to which, for example, content moderators find themselves exposed when cleaning up platforms such as Facebook or YouTube from violent depictions and hate speeches (Roberts, 2019). On the other hand, critical social science studies are increasingly identifying common principles and patterns that underlie as well as emerge from

---

2 The term was coined by Amazon founder Jeff Bezos during an opening keynote at MIT Emerging Technologies Conference in 2006 (MIT 2018).

3 Silicon Valley solutionism refers to prevailing ideology and beliefs that for each and every social challenge there is technical solution (see Techrepublic 2014).

4 "The Turk" or "Mechanical Turk" was a eighteenth century mechanical man dressed in a Turkish costume, that was capable of playing chess. While the set-up was to pretend that the Turk could play automatically it was actually fake, since its chess skills relied on a human hidden inside the machine (Standage 2002).

specific labour conditions and working practices considered increasingly characteristic to the micro, gig, or crowdsourcing economy (Vallas and Schor 2020). According to these studies, most forms of employment are built on exploitative labour conditions with low payment (Fish and Srinivasan 2011), and hardly any labour law security (Ettinger 2017), which from a global perspective contributes to the emergence of a new digital precariat (Irani and Silberman 2013).

However, case studies that explore the current and future role and relevance of the internet for this new young field of globally circulating labour address, to varying degrees, the issue of unequal distribution of access to digital services and employment opportunities. One concept that attempts to capture the issue of unequal access is the notion of a growing digital divide (Mariscal 2005, Fuchs and Horak 2008). A central observation of this concept is that an increase in access to the internet and digital services is simultaneously accompanied by an increase in digital disconnectedness of large parts of the world's population (World Bank 2016). Despite the World Bank's concerns regarding growing global inequalities, its framing has been criticized as a quest for yet another technical fix (Donner 2015, Shrum et al. 2007). If such a framing would be effective, a purely technical availability of the internet would have beneficial and accelerating impact on the economic development of a country or region.

An argument that this does not have to be case is provided by Louise Bezuidenhout et al. (Bezuidenhout et al. 2016, Bezuidenhout et al. 2017) showing that the mere availability and access to opensource and open-data science in the context of African universities depends on further requirements and skills in order to unfold positive impact for academic research. The question of when the internet can actually serve as a resource for an individual or a group to cope more efficiently and successfully with the hardships of everyday work is not simply solved by providing access. While I consider this perspective to be valuable and share the associated criticism of the concept of the digital divide, I will propose a perspective that is diametrically opposed to the argumentation of Bezuidenhout. For this, I will argue that we currently find particular forms of digital work—and uses of the internet—whose functioning builds and relies on the exclusion of, and disconnection from, social, material, personal, and knowledge-based requirements. The planning, logistics, organisation, and core functional/technical principles of digital wage labour in African contexts, for example, are feasible and economically successful, precisely because they disconnect the enterprise from local entanglements and situated knowledge cultures. The effects as well as their side effects will be presented in the following empirical sections.

3   Material and Infrastructural (Dis)Connections

The research project that allowed me to travel to Gulu again in September 2018 was dedicated to the development and expansion of university curricula.[5] One initial aim was to introduce social science perspectives on science and technology as part of humanities and engineering courses. Furthermore, the project also aimed to augment teaching capacities through the dissemination of teaching materials from online courses such as Massive Open Online Courses (MOOCs). At this point during the project, my involvement with the university's internal information and communication infrastructures (ICTs) intensified, as it quickly became clear that only functioning and reliable ICTs would enable the provision and use of e-learning formats at all. During a meeting with the dean of the Computer Science Faculty, I asked where he would be getting internet access from. While answering my question, he pointed to a container just outside his window and added that he would get good internet connection from Samasource: "Theirs is more reliable and faster than the one we have on campus."[6] Indeed, shortly before, I had been told by the university's network manager that the campus Wi-Fi connection was very limited. In order to better understand this parallel and simultaneous functionality and non-functionality of internet connections, it is necessary to look into the more recent history of the northern Ugandan region.

Until 2008, the municipality as well as the surrounding district of Gulu formed the centre of civil war that started in the early 1980s and lasted for decades (Büscher et al. 2018). From my first visit in 2010 until my last stay in May 2019, Gulu has increasingly transformed into a regional commercial centre connecting several important trading arteries. During this time, the city became a hub for international non-governmental organisations whose post-conflict development projects not only brought far-reaching financial resources, but additionally offered new diverse employment opportunities to the region (Redfield 2013). Even until today, most major aid organisations (e.g. UNICEF, Oxfam, USAID, GIZ, MSF) continue to have regional offices in the city. In recent years, Gulu has also served as logistical hub for other crisis

---

5   The project funded by the German Research Foundation (DFG) entitled Curriculum Development (2017–19) was based at the Department of Social Anthropology at Martin Luther University Halle (see Lost n.d). During the field visits of three weeks each (September 2018 and May 2019), I conducted three focus group discussions (University staff, students, Samasource staff) and ten semi-structured interviews with current and former Samasource staff in Gulu. Other quotations and information are taken from sporadic conversations and chance encounters on Gulu campus between Samasource management and myself.

6   Interview #1, Dean of Faculty Computer Science, Gulu, September 18, 2018.

regions. Currently, United Nations Refugee Agency (UNHCR) and the World Food Programme (WFP) coordinate parts of the humanitarian aid efforts in South Sudan from here. This connectedness to the international donor community is also mirrored throughout the university's faculties, whose individual departments are often part of international research collaborations with mainly European and American universities.

As part of its post-conflict recovery and development measures, Oxfam—a UK-based, non-governmental organisation—initiated a supra-regional project called *Internet Now!*. Starting in 2012 the aim was to make internet-based learning and employment options available focusing largely on rural areas. In collaboration with the Gulu University administration, ten containers were placed on campus providing equipment and resources for approximately thirty computer science students and graduates; allowing them to carry out early forms of micro-work. After initial collaborations with Oxfam, Samasource took over the Gulu branch of Internet Now! in 2016.[7] Samasource is one of the pioneering organisations in the field of impact sourcing, a subtype of Business Process Outsourcing (BPO), a widespread corporate practice through which sub-processes and tasks are outsourced to other specialised companies. However, unlike BPO, impact sourcing companies claim that the services they provide are managed through employment of marginalised and disadvantaged populations (Heeks 2013). Against this background Samasource promises to address two problems at once: "the big data problem and the global poverty problem" (Janah 2017, 118). According to the organisation's self-portrait, Samasource offers jobs and training for digital activities to people below the poverty line in Kenya, Uganda, Haiti, and India. Since its foundation in 2008, Samasource has launched sixteen branches—of which some are no longer operational—in which it employed an estimated eight thousand workers. The organisation's motto "Give work, not aid" (Janah 2017, 3) follows a free market logic, with its core assumption that the provision of wage labour opportunities empowers individuals in developing countries—better than, for instance, international aid—to lift themselves and their families out of poverty. The extent to which impact sourcing attempts can actually contribute to the reduction of global poverty is difficult to answer.

For Samasource Gulu, one can assert that it is predominantly students and graduates who are attracted to and recruited by the organisation.[8] Even

---

7   Unfortunately, I was not able to piece together more precisely which exact circumstances led to both the initial collaboration and the eventual takeover by Samasource.

8   It is difficult to answer which tiers of society Samasource mainly recruits its workforce from. However, in locations like Nairobi, Samasource does not operate directly on a university campus, which might potentially have an impact on who is applying and eventually recruited.

though a large number of students come from less affluent backgrounds—for instance, over 50 percent of computer science students do not own a computer during their first or second year of study—they likely do not belong to the very poor bottom echelons of society. Studies on the actual effects of impact sourcing have also plausibly pointed out that the social impact of outsourcing bulk work plays only a secondary and subordinate role to many companies. "Executives like hearing about the company's social mission, but the decision to hire the organization is made by purchasing departments, which have to think about cost and quality. And many buyers harbour preconceived notions that non-profits are slow and uneconomical" (Gino and Staats 2012).

In order to stay operational in such a heavily competitive field of business outsourcing, Samasource Gulu must guarantee a high level of quality and reliability, starting at the technical and infrastructural level. Running an outpost like Gulu exposes their operations to heavy and long power cuts often lasting several hours a day.[9] After the takeover of *Internet Now!* Samasource added sixteen additional containers and made room for one hundred more working stations. To sustain this extension a new generator had to be purchased, as well as a second internet connection, which now serves as a back-up in case the first connection is interrupted. William, the manager summarises the situation as follows: "We run our own system here! It is completely independent from the university. If we had to rely on their system we would definitely fail."[10] Even though there are only a few metres between the university's own computer lab and Samasource containers, the contrast could not be greater. Extreme heat, high humidity, and fine dust have turned the computer lab into an unused, sad museum offering an unintended glimpse into the history of computer technology. Especially during power outages after sunset, the brightly lit windows of the Samasource containers are often the only light on the entire campus; a light that denotes productivity, but a productivity that is fully decoupled from local realities and spaces. The costs of these material and infrastructural decoupling measures are considerable. The internet connection alone amounts to a cost of at least $20,000 per year. It should be clear by now that the conditions for an organisation like Samasource to have settled in a location like Gulu are fundamentally related to the costly option of disconnecting its operations from the local daily material and infrastructural uncertainties.

---

9   During these power cuts, a campus generator can be switched on to supply power to the rest of the campus. However, the cost of diesel consumption has to be paid for from university funds.

10   Interview #4, Manager Samasource, Gulu, May 6, 2019.

4   Knowledge Decoupling: Logistical Media and Workplace Studies

"Sometimes I find myself annotating faces of people I interact with."[11] This statement was made by a participant during a focus group discussion in which interlocutors spoke about the social effects of annotation activities. Other participants also remarked on its psychological effect, which extended into leisure time, and they attributed it to the predominantly repetitive nature of most of the activities. In general, most tasks at Samasource were described as monotonous, boring, and tiring. However, performing these tasks six days a week, for at least ten hours a day, forms the organisational basis and inner logic of data factory work. Samasource founder Leila Janah,[12] was inspired by Henry Ford's assembly line when developing her micro-work principles: "Ford figured out a way to break down the making of an incredibly complex machine [the Ford Model T, RU] into small chunks that people with basic training could complete. He moved the Model T from the craftsman's studio into the mainstream. The assembly lines of the future apply the same thinking to digital work." (ITWeb, 2021). Whereas in Ford's factories it was the assembly line that dictated the rhythm and intensity of work, today the interface between employees and their workload is mediated by a more dynamic software. At Samasource this function is performed by SamaHub, its own in-house digital platform that has been developed over years with the following core functions: "SamaHub is Samasource's proprietary training data annotation platform. This web-based task management system helps facilitate large data projects through the customization of task workflows, task distribution, multi-tiered quality control, and project-based training. [...] SamaHub has three main functions: task distribution, data structuring work (image annotation, data categorizing, other writing/research) and quality management" (Samasource 2019, 1).

Terms such as "task-decomposition," "task-distribution," or "task-completion" indicate sub-steps in which the software automatically splits up projects (e.g., large files of visual data), distributes it, and also returns it to clients and customers as valuable annotated training data. This has become a fairly complex operation in which clients have the option to check and control the data labelling in real-time. They can change data volumes, and also make use of the labelled data almost immediately. The activities indicated in Figures 4.1 and 4.2

---

11   Focus Group Discussion #5, Samasource Staff, Gulu, May 5, 2019.
12   Leila Janah passed away on 24 January 2020. Until then, she ran the cosmetic label LXMI in addition to Samasource. The label primarily sells products that use shea butter as a key ingredient. The nuts form the shea trees were also harvested in Uganda, and Janah also promoted this branch of the company as providing marginalised women with meaningful and sustainable employment (see LXMI 2021).

FIGURE 4.1   Face annotation (Edge AI and Vision Alliance 2019a)
EDGE AI AND VISION ALLIANCE. (2019A). SAMASOURCE DEMONSTRATION OF
ITS SAMAHUB TRAINING DATA ANNOTATION PLATFORM. YOUTUBE VIDEO,
2:57. LAST MODIFIED OCTOBER 9, 2019. HTTPS://WWW.YOUTUBE.COM
/WATCH?V=FDS09IUAGMO

FIGURE 4.2   Geographical area annotation (Edge AI and Vision Alliance 2019b)
EDGE AI AND VISION ALLIANCE. (2019B). SAMASOURCE DEMONSTRATION OF
ITS SAMAHUB TRAINING DATA ANNOTATION PLATFORM. YOUTUBE VIDEO,
2:57. LAST MODIFIED OCTOBER 9, 2019. HTTPS://WWW.YOUTUBE.COM
/WATCH?V=FDS09IUAGMO

demonstrate what most of tasks at Samasource consist of. Only through
additional labelling of individual images, or entire videos, does "raw-data"
becomes useful for learning algorithms. One could also say that these enrich-
ment practices that transform data into training data constitute the "Artificial"
in Artificial Intelligence. Each point between the lines must be set manually
by a mouse click and then saved under one of the pre-formatted categories
or labels (e.g. nose, eyebrows, livestock, house). The tracing and drawing of
shapes, objects, or body parts is based on simple phenomenological identification that is assumed to work without reference to cultural standards or values and thus without much interpretative effort.[13] In this way annotations can
travel globally, creating a work realm that is largely acultural with extremely

---

13   When I write here "without much interpretative effort", I am nevertheless aware that it is
       up to the interpretation of an individual, for example, where a house exactly begins and
       ends. These activities carry less moral interpretation than tasks assigned to the already
       mentioned content moderators. The latter must be able to assess throughout several
       cultural, aesthetic, religious and ethical codes if something is considered violent or hate
       speech. See chapter by Helen Robertson in this volume.

limited interpretative requirements. Even without prior experience of digital data processing, lay persons can perform most activities with minimal training and regardless of their geographical origin. As the initial statement indicates, the continuous and repetitive nature of the tasks can still have some social or psychological impact on employees' imaginations and perceptions of bodies, as well as faces and infrastructures, which I however did not follow up further in this research (e.g., Bajwa et al. 2018).

The comparison with Ford's assembly line can also be found within the execution speed and the timing of individual labelling tasks. For instance, the average time an employee has for annotating a face is usually calculated by the SamaHub platform. With each new face the time is displayed as a countdown above the image. Next to the countdown called *default time*, is another timer counting up the real time that is called *elapsed time* that the employee actually needs for fulfilling the task. This temporal regime of (self)control affects the performability of the respective activity in several ways. At the individual level, performance that is measured as too slow will negatively impact the employee's internal evaluation, which again affects the future tasks they will be assigned to, as well as employment and contractual negotiations. On a more practical level, if SamaHub detects that the default time is collectively exceeded, adjustments can be made that eventually result in an extension of default time. A helpful conceptualisation of this partial operational logic of labour control through digital platforms is called "logistical media theory" (Rossiter 2016, 7): Logistical labour emerges at the interface between infrastructure, software protocols, and design. Labour time is real-time. The formation of logistical media theory, therefore, requires an analysis of how labour is organised and governed through software interfaces and media technologies that manage what anthropologist Anna Tsing (2009) identifies as "supply chain capitalism." (Rossiter 2016, 23)

This logistical extraction of labour is thus based on a twofold decoupling: On the one hand the breaking down of videos into individual frames decontextualizes the content, as it operates with a simplification that allows a decoupling from any local knowledge practices. On the other hand, the software offers a significant decoupling from local, often limited, skillsets and work experiences. Even without any IT skills people can easily be enrolled and made part of this new form of digital value production and supply chain capitalism.

Analogous to Ford's workers at the assembly line, employees of Samasource also attempted to momentarily escape control, monitoring, and supervision.[14] The management used the term "outliers" indicating that they are aware of

14   However, I would not go as far as to call these practices acts of sabotage.

some of the informal appropriation practices by their employees. Almost all employees who were interviewed admitted that they knew, for instance, how to trick SamaHub's strict temporal regimes. However, only few admitted that they engaged in such practices while the majority said they refrained from them most of the time because they feared being punished or discharged. In the words of the manager William: "Of course we have outliers that crack the system. What do you expect? These are young smart people. First thing they find out how to circumvent and it takes some time for us to find out and correct it ..."[15]

5   Decoupling from Local Work Cultures

Calculatory practices that relate the scheduled amount of labour to the actual labour needed is part of broader history of organisational learning and indus-trial labour culture, which in the case of Samasource Gulu, interacts with but also conflicts with, local work practices. This final section will be dedicated to the discussion of the relation between digital factory labour requirements and local labour cultures. Apart from a few companies in the capital city of Kampala, elaborate control and tight time keeping devices are rarely part of the everyday work experience of most Ugandans. In addition to the entrance login, and the login at their workplace, all employees of Samasource have to hand in their mobile phones on entering the building. Mobile phones may only be used during strictly recorded breaks. Most of the interviewees found the mobile phone policy to be the most unpleasant control measure. It was repeat-edly indicated that during working hours, employees could not respond, for example, to urgent family needs and other issues.[16] It also regularly happened that individual workers completed their daily workload well before the end of their shift. I heard complaints from employees indicating that after having fin-ished early, for example, around three o'clock in the afternoon, they were nei-ther allowed to leave nor could they use the internet for private use until the end of the shift, usually at six o'clock in the evening. One employee described the situation as follows: "When you finish early you are just supposed to sit and do nothing."[17] When I asked the management about their policy on that, I was told that there were indeed waiting times every now and then, particu-larly because quality control managers had to check their team members work

---

15   Interview #4, Manager Samasource, Gulu, May 6, 2019.
16   In this context, it was also mentioned that young mothers were probably not allowed to breastfeed babies on the company premises during working hours.
17   Focus Group Discussion #3, Samasource Staff, Gulu, September 19, 2018.

which resulted in backlogs sometimes.[18] In principle, however, I was assured that after finishing their daily bulk, employees could use the internet or that an earlier end of shift could be arranged.

On the one hand the formalisation techniques for labour registration and worktime measures practised at Samasource did not go beyond corporate governance policies usually found in transnationally operating companies. On the other hand, the above-described contradictions regarding the end of shift also suggest that strict formal regulations interact with local forms of social control within corporate contexts. In the words of one informant: "The problem is not Samasource. The problem is on the ground."[19] The reflection refers to patronage relationships among the personnel through which access to tasks and resources are either granted or restricted. Starting from recruitment and allocation of shifts (day/night), kinship networks, and even ethnicity formed dominant inclusion or exclusion criteria potentially undermining merit principles. Another complaint was that it mattered whether one was favoured by middle management. When I enquired why the management in San Francisco headquarters did not attempt to regulate certain abuses of power, it was mentioned that they were largely kept in the dark about the local conditions. During visits from the San Francisco office, only pre-selected employees who were considered loyal were allowed as contacts.

In light of these social and organisational entanglements, it remains a challenge to judge the extent to which this form of repetitive and uncreative data work is adequately renumerated. During my fieldwork, however, this question repeatedly emerged in a strangely paradoxical manner: When I reported the type of labour practiced at Samasource Gulu to Ugandan colleagues, they rarely shared my critical questioning of the monthly payments. Instead, an income of USh460,000 per month (approximately $120) was deemed fair pay for a university graduate. In the broader national context such income of newly graduated students would sit closer to around USh250,000 (approximately $65) for

---

18   The issue of quality data has gained some importance as it serves as a niche for companies like Samasource, which helps them justify and distinguish their costly and elaborate factory-style operations against the remote and individualised operational characteristic of many other sectors of the gig economy. IT-Tech as well as automotive companies require properly labelled data for their learning algorithms that are often only achieved through strict monitoring and quality control measures. Sending out micro-tasks on crowd platforms, like AMT or Upwork, exposes the data to significant differences in the quality that is produced. See in this regard also Anwar and Graham's (2019) study on adaptation practices of African gig workers to cope with the (labour) conditions they are facing on Upwork.

19   Focus Group Discussion #5, Samasource Staff, Gulu, May 5, 2019.

a month of work. In an interview with the BBC, founder Leilah Janah justified this salary policy as follows: "One thing that's critical in our line of work is to not pay wages that would distort local labour markets. If we were to pay people substantially more than that, we would throw everything off. That would have a potentially negative impact on the cost of housing, the cost of food in the communities in which our workers thrive" (Lee 2018).

Although this justification seems to be rather speculative and less grounded in empirical evidence, at this point I would like to draw attention to a specific feature of Ugandan work culture, which is not taken into consideration in Janah's statement. In Uganda, as well as in other countries with comparatively low incomes, the line between formal wage labour and informal employment remains a difficult and complex distinction. However, through my research in the Ugandan health sector, I became aware that the distinction between the state/public and the private sector plays a significant role in perception of local work cultures (Umlauf 2017). For example, it is a very likely scenario that a nurse or a doctor employed in the private sector would receive a higher monthly income than in the same job in the public sector. Nevertheless, a large proportion of private sector health workers would prefer to work in the public sector. The main reason, apart from generally higher job security, was the opportunity to participate in further training sessions. As part of internationally funded donor projects, these capacity building missions take place several times a year and are highly attractive, especially because of their daily allowances, which often bring in several additional monthly salaries (Swidler and Watkins 2009). In addition to these benefits of accessing the development industry, almost all government employees were engaged in side businesses constituting an additional source of income (McCoy et al. 2008). In contrast, greater control and regulation of working hours of privately employed health workers make it much more difficult to operate and sustain secondary incomes. The strict labour regulations and control of working hours at Samasource, together with the strongly limited communication options, conflict in various ways with local working cultures and the need to divert sources of income. Against this background a higher pay could thus be justified relatively easily without fear of distorting local wage and labour markets, as indicated in Janah's statement.

I would like to conclude this section by returning to Samasource's ethical claim of lifting people out of poverty by giving them jobs instead of aid. The fact that marginalised tiers of the population are mainly characterised by poor access to formally secured wage labour is reinforced by the aspect that many have only limited knowledge of labour protection and labour law regulations. As already mentioned, we can assume that the majority of employees at Samasource Gulu do not belong to the poorest or most marginalised echelons of Ugandan society. Nevertheless, these are mainly young, recent graduates in their first official employment roles. Combined with the prevalence of patronage relations mentioned above, the inexperience of most employees prevents substantial articulation of collective forms of resistance. The paucity of collective action is a feature of the crowdsourcing industry as a whole, and is related to the difficulty of a globally dispersed crowd to meaningfully and impactfully unite in order to claim recognition and negotiate better working conditions and fair payment (Ettlinger 2016). In this respect, the case study of Samasource shows that even for centralised, factory-style mass activities, it seems difficult to negotiate more favourable working conditions or make collective claims for higher salaries, which could better compensate for the limiting of access to and caring for additional sources of income. Graduates-as-a-service allows companies mainly from the United States, Germany, and China to save money—as the price of having this kind of work done in their countries of origin would be significantly higher—by employing and relying on an energetic but still inexperienced workforce.

6   Conclusion

Entitled "Digital Dividend," the 2016 World Bank report states that, in a historical comparison, the internet reached many developing countries much faster than was the case for other technologies (World Bank 2016, 5). At the same time and despite the spread of mobile money and the use of social media, the everyday lives of the majority of people living in the so-called developing world still offers very limited interfaces with digital services and their promised benefits. With reference to these observed frictions, I have attempted to outline a more complex narrative of possible and actual forms in which access to digital services and employment opportunities are currently (infra)structured in Uganda. The ethnographic case study of Samasource Gulu was intended to demonstrate the ambivalence with which the spread of the internet takes place. On the one hand, it extends largely to very specific and strongly regulated sub-sectors, such as the digital mass and micro-labour. On the other hand, the study was also meant to illustrate how software-driven tailoring of the internet neither increases digital literacy nor does it produce much value for other parts of people's everyday lives. I have shown that the connection and access to new digital employment opportunities both requires and goes hand in hand with disconnections from other forms of local knowledge systems and work cultures.

It remains difficult to predict what future developments will emerge from this still nascent form of digital wage labour in low-income countries. At the present stage—but probably also in the near future—a stable and even increasing demand for inexpensive training data for machine learning and AI applications is likely to persist. At the same time, there are indications of trends within the digital economy in which for example, neural networks are either no longer dependent on this kind of enriched data (e.g., self-learning algorithms) or automated applications—again based on algorithms—are being developed. These greatly reduce the demand for humans-as-a-service or make them redundant all together.[20] For smaller and more remote sites like Gulu, the latter prognosis means an uncertain and unsustainable future scenario. While the containers can be quickly dismantled or repurposed according to their logistical nature, much more uncertainty prevails according to where or which employment sector the approximately three hundred graduates could be absorbed. The skills learned at Samasource are not very specific and hard to apply beyond the organisations own interface (SamaHub). It can also turn out to be problematic that long-term employment at Samasource does not translate into any professionalisation within the originally studied field of occupation for most graduates, but—on the contrary—can potentially have negative effects on their future employment prospects.

Bibliography

Alexander, Jan. 2018. "How Samasource Reverses Poverty One Job at a Time." *This is Capitalism.* Last modified January 8, 2018. https://www.thisiscapitalism.com/samasource-reverses-poverty-one-job-time/.

Anwar, Mohammed Amir and Mark Graham. 2019. "Hidden Transcripts of the Gig Economy: Labour Agency and the New Art of Resistance Among African Gig Workers." *Environment and Planning: Economy and Space* 52, no. 7: 1–23.

Bajwa, Uttam, Denise Gastaldo, Erica Di Ruggiero, and Lilian Knorr. 2018. "The Health of Workers in the Global Gig Economy." *Globalization and Health* 14, no. 124.

Bergvall-Kåreborn, Birgitta and Debra Howcroft. 2014. "Amazon Mechanical Turk and the Commodification of Labour." *New Technology, Work and Employment* 29, no. 3: 213–23.

---

20   The threat of potential redundance of this field of employment is already in the minds of the Samasource management: "In spite of such success stories, however, Janah is frank about the challenges ahead. The greatest concern is that as technology continues to advance, in five to ten years machines might perform the tasks that Samasource is now training people to do" (Alexander n.d).

Bezuidenhout, Louise, Ann Kelly, Sabina Leonelli, and Brian Rappert. 2016. "'$100 Is Not Much To You': Open Science and Neglected Accessibilities for Scientific Research in Africa." *Critical Public Health* 27, no. 1: 39–49.

Bezuidenhout, Louise, Sabina Leonelli, Ann Kelly, and Brian Rappert. 2017. "Beyond the Digital Divide: Towards a Situated Approach to Open Data." *Science and Public Policy* 44, no. 4: 464–75.

Büscher, Karen, Sophie Komujuni, and Ivan Ashaba. 2018. "Humanitarian Urbanism in a Post-Conflict Aid Town: Aid Agencies and Urbanization in Gulu, Northern Uganda." *Journal of Eastern African Studies* 12, no. 2: 348–66.

Daub, Adrian. 2020. *What Tech Calls Thinking: An Inquiry into the Intellectual Bedrock of Silicon Valley.* New York: Farrar, Straus and Giroux.

Donner, Jonathan. 2015. *After Access: Inclusion, Development, and a More Mobile Internet.* Cambridge, MA: MIT Press.

Ettlinger, Nancy. 2016. "The Governance of Crowdsourcing: Rationalities of the New Exploitation. *Environment and Planning A: Economy and Space* 48, no. 11: 2162–80.

Ettlinger, Nancy. 2017. "Paradoxes, Problems and Potentialities of Online Work Platforms." *Work Organisation, Labour & Globalisation* 11, no. 2: 21–38.

Fish, Adam and Ramesh Srinivasan. 2011. "Digital Labour Is the New Killer App." *New Media & Society* 14, no. 1: 137–52.

Fuchs, Christian and Eva Hora. 2008. "Africa and the Digital Divide." *Telematics and Informatics* 25, no. 2: 99–116.

Gino, Francesca and Bradley R. Staats. 2011. "Samasource: Give Work, Not Aid." Harvard Business School Case 912–011, December.

Gray, Mary and Siddharth Suri. 2019. *Ghost Work: How to stop Silicon Valley from Building a New Global Underclass.* Boston: Houghton Mifflin Harcourt.

Heeks, Richard. 2013. "Information Technology Impact Sourcing." *Communications of the ACM* 56, no. 12: 22–25.

Irani, Lilly. 2015. "The Cultural Work of Microwork." *New Media & Society* 17, no.5: 720–739.

Irani, Lilly and Six Silberman. 2013. "Turkopticon: Interrupting Worker Invisibility in Amazon Mechanical Turk." *CHI 13: Proceedings of the SIGCHI Conference on Human Factors in Computing Systems*, 611–20. Paris.

ITWeb. n.d. "Business Technology News." Accessed January 11, 2021. https://www.itweb.co.za/content/mYZRXM9JyD8qOgA8.

Janah, Leila. 2017. *Give Work: Reversing Poverty One Job at a Time.* New York: Penguin.

Lee, Dave. 2018. "Why Big Tech Pays Poor Kenyans to Teach Self-Driving Cars." BBC News. Last modified, November 3, 2018. https://www.bbc.com/news/technology-46055595.

Lost Research Group. n.d. *Lost Research Group.* Accessed November 16, 2021. https://lost-research-group.org/.

LXMI. n.d. "Ultra Clean Skincare | Organic & Fair Trade | LXMI". Accessed January 10, 2021. https://lxmi.com/.

Mariscal, Judith. 2005. "Digital Divide in a Developing Country." *Telecommunications Policy* 29, no. 5–6: 409–28.

McCoy, David, Barbara McPake, and Sara Bennett. 2008. "Salaries and Incomes of Health Workers in Sub-Saharan Africa." *The Lancet* 371, no. 9613: 675–81.

MIT ODL Video Services. 2006. *Opening Keynote and Keynote Interview with Jeff Bezos*. MIT ODL video, 1:02:38. Last modified December 4, 2018. https://techtv.mit.edu/videos/16180-opening-keynote-and-keynote-interview-with-jeff-bezos.

Redfield, Peter. 2013. *Life in Crisis: The Ethical Journey of Doctors Without Borders*. Berkeley: University of California Press.

Roberts, Sarah. 2019. *Behind the Screen: Content Moderation in the Shadows of Social Media*. New Haven: Yale University Press.

Rossiter, Ned. 2016. *Software, Infrastructure, Labor: A Media Theory of Logistical Nightmares*. New York: Routledge.

Samasource. 2019. *Samahub Process and Feature Overview*. Samasource. Accessed January 10, 2021. https://www.samasource.com/hubfs/Collateral/SamaHub%20Solution%20Brief%202019.pdf.

Selwyn, Neil. 2004. "Reconsidering Political and Popular Understandings of the Digital Divide." *New Media and Society* 6, no. 3: 341–62.

Selwyn, Neil and Kirk Facer. 2010. "Beyond Digital Divide: Toward an Agenda for Change." *Handbook of Overcoming Digital Divides. Constructing an Equitable and Competitive Information Society, Volume 1*, edited by Enrico Ferro, Yogesh Kumar Dwivedi, J. Ramon Gil-Garcia, and Michael D. Willaims, 1–20. Hershey, PA: IGI Global.

Shrum, Wesley, Keith Benson, Wiebe Bijker, and Klaus Brunnstein. 2007. *Past, Present and Future of Research in the Information Society*. New York: Springer.

Standage, Tom. 2003. *The Turk: Life and Times of the Famous 18th Century Chess-Playing Machine*. New York: Walker & Co.

Swidler, Ann and Susan Cotts Watkins. 2009. "'Teach a Man to Fish': The Doctrine of Sustainability and Its Effects on Three Strata of Malawian Society." *World Development* 37, no. 7: 1182–96.

Techrepublic. 2014. "Silicon Valley's 'Solutionism' Issues Appear to be Scaling." Last modified August 25, 2015. https://www.techrepublic.com/article/silicon-valleys-solutionism-issues-appear-to-be-scaling.

Umlauf, René. 2017. *Mobile Labore. Zur Diagnose und Organisation von Malaria in Uganda*. Bielefeld: Transcript Verlag.

Vallas, Steven and Juliet B. Schor. 2020. "What Platforms do? Understanding the Gig Economy." *Annual Review of Sociology* 46, no. 5: 273–94.

World Bank. 2016. *World Development Report 2016: Digital Dividends*. Washington D.C.: World Bank Publications.

CHAPTER 5

On the Technopolitics and Metrics of the Right to Information

Jonathan Klaaren

1 Introduction

This chapter explores the South African experience with the justiciable right of access to information introduced by the first democratic constitution in 1994. Written initially as part of section 23 of the 1994 Constitution and then as section 32 of the 1996 Constitution, the words of this right have no reference to the technological media of information. Adopting an STS-inspired perspective, the chapter particularly focuses on how the right of information emerged entangled some ten years later with the justiciable right to privacy. The implementation of both rights is related to the rapid increase of digitisation and the ensuing surge of digitised quantitative information. One question to ask is how did this legal discourse shift its focus from looking at the right of access to information to framing the technology of digitisation as a means to achieving the social and democratic goals of the Constitution. Did this shift result in a change of purpose? Could one say it represents the organised reactive political triumph of forces of bureaucracy, private firms, elites, and technocrats against what could have been a breakthrough of popular participation and bottom-up democracy?

With this general question in mind, this chapter explores a more specific problem posed by the introduction into the South African legal order in April 1994 of a justiciable right of access to information, a problem with two distinct aspects. A first problematic—one common to all legal rights—is that the right of access to information is by no means a self-executing fact. The right of access to information was not an immediately realised idea on 27 April 1994 when the interim Constitution came into legal effect. The right was justiciable but it was not fulfilled. As the term "justiciable" implies, the right was able to be adjudicated upon—for instance called into force through a court-proceeding—but the mere fact that a court now had jurisdiction over this right did not change the practices of most (if any) government departments overnight.

As has been recognised by a diverse, detailed, and exciting body of interdisciplinary scholarship, law on the books is not the law in action (Calavita 2010).

Likewise, rights in the air are not rights on the ground, until some concrete action—whether by a court, an organisation, or another social actor—is taken (Rosenberg 2008; Robins 2008). Nonetheless, rights can give rise to hope and to ways of imagining and communicating a better life (McCann 1994; Dugard 2008). Rights have social lives.

We shall return to this first fundamental aspect of the problem posed by the right of access to information further below in this chapter. But immediately upon its introduction (and indeed in its social life prior to its South African birth), a second problematic of the right of access to information also became apparent—the legal and social relationship of the right of access to information with the right to privacy. Discussed further below, this problematic between the access to information right and the right to privacy drives much of the story of this chapter. Indeed, there is perhaps an even deeper problematic here. At least in fundamental part, these two rights—access to information and privacy—rest on the concepts of openness and secrecy. The inherent contradiction between openness and secrecy may be seen to be straightforward: the realisation of the one goal (openness) tends to suspend the realisation of the other goal (secrecy), and vice versa. However, the contradiction is a dialectic one. Secrecy is a kind of knowledge that opposes but also supplements transparency (Horn 2011; Nuttall and Mbembe 2015). In this vein, the openness that was introduced to South Africa as part of its democratic transition can only be understood together and simultaneously with secrecy (Klaaren 2015). Transparency only makes sense by understanding it to be entangled with opacity and by investigating it in specific historical and cultural contexts (Birchall 2011; Pozen 2018). As the goals of openness and secrecy remain important and are inherently related, their realisation can best be achieved in cognisance of this relation. Recognising the dialectics between these goals, the chapter elaborates how legal attempts to pursue the rights of access to information and privacy have co-produced a sociopolitical order together with three information technologies.

The conjoined legal twin of the right to access to information was the right to privacy, a legal right which also entered into (justiciable) force at the constitutional level in the South African legal order only on 27 April 1994. Both rights were recognised and shaped within the specific imaginary of open democracy associated with South Africa's democratic transition—a vision reacting sharply against apartheid secrecy. There was however at least one significant legal and social difference between these two rights. The right to privacy had a previous life in apartheid. Legally speaking, South Africa had a common law tradition of a right to privacy to draw upon when its constitutional right to privacy was introduced (Cachalia and Klaaren 2022). Perhaps not surprisingly this non-constitutional right to privacy extended to and was used by corporations. The constitutional right of access to information had no legal predecessor in the apartheid era. It thus had a clean slate, no legal infrastructure of court cases to build upon, although there was at least a faint liberal tradition of invoking transparency to counter apartheid secrecy (Bozzoli 1975; Mathews 1978). A second important difference soon emerged. While the right of access to information was fairly quickly implemented in the form of national legislation,—the Promotion of Access to Information Act 2 of 2000 (PAIA)—the right of privacy would wait thirteen further years for its national law—the Protection of Personal Information Act 4 of 2013 (POPIA)—to be drafted and enacted onto the statute books.

This chapter approaches its subject matter from the perspective of openness limited by confidentiality, not from the perspective of secrecy eroded by transparency. The chapter focuses on how the social life of the twin statutes—PAIA and POPIA—unfolds within or in relation to public administration. By public administration, this chapter means what one might term the organisational infrastructure of the South African state: the government departments and other public organisations that to a great extent make the state a real force in the lives of persons living in (and beyond) the borders of the South African territory. As these organisations are uneven (Klaaren 2021; Naidoo 2019; Brunette 2014) they ideally would require deeper ethnographic and historical analysis beyond the scope of this chapter. Instead, this chapter will focus upon the two problematics identified above. At one point in the story told below, the linkage of the right of access to information to the privacy right bursts upon the scene.

Returning now to the introduction of the right of access to information and the unavoidable fact of its non-automatic implementation, the aim of this chapter is to inquire into the co-production of the technical and legal order that happened after 27 April 1994. This co-production of order occurred between the PAIA and its legal interpretation and application on the one hand and the creation of order through information technology by the government departments that the PAIA drafters saw as their primary targets and as the chief bearers of the duties imposed by the right. There are (as of the time of finalising this paragraph) forty-six national government departments subject to the PAIA as well as greater numbers of provincial departments and of local government entities. However, for the particular purpose of this chapter it is in order to construct an ideal type government department that will stand in for all of them. The myriad of variations from department to department are not necessary to show how the construction, use, and interpretation of the PAIA was a social practice that interacted with the construction, resourcing, and operation of the typical South African government department. This interaction produced the social order that will be described in this chapter as the social life of the right of access to information. Additionally, that order was co-produced through technical and legal developments occurring both on the terrain of the PAIA (and POPIA) and on the more bureaucratic terrain of government departments from 1994 to the present.

In the way this chapter problematises its object of study, it draws upon the work of Sheila Jasanoff who is a foundational scholar of science and technology studies and stands out by her attention to the legal dimensions of science and technology (Jasanoff 2009). Accounts researched and written in a co-productionist framework often investigate four "pathways" identified by Jasanoff: making identities, making institutions, making discourses, and making representations (Jasanoff 2004). This chapter will mainly draw upon two of Jasanoff's pathways in exploring its story below: institutions and discourses.[1] Within a context of changing information technology (broadly put, the growing influence of digital technology over the twenty-five-year span examined) how have actors tried to put into action the right of access to information and how has that effort been resisted in part by officials of the public administration? Specifically, how has the discourse of openness changed—have new aspects emerged? And to what extent were old institutions (e.g. the public bodies of the South African state) remade or not remade and to what extent were effective new institutions built?

Some literature touching upon constitutional rights and their realisation in postapartheid South Africa has focused on the entanglements of law, science, technology, and their political translations into issues of administration and service provision (Plantinga, Adams, and Parker 2019; Plantinga and Adams 2021; Ravigopal 2019; Mpofu-Walsh 2021, 106-32). One excellent example is Laura Foster's account of the postapartheid saga of intellectual property in a plant used by indigenous people, hoodia (Foster 2017). Foster explores how the plant was reinvented through patent ownership, pharmaceutical research, the self-determination efforts of indigenous San peoples, contractual benefit sharing, commercial development as an herbal supplement, and bioprospecting legislation. And she explores how the plant helped different communities including scientists and the San indigenous peoples to form and make claims for belonging within South and Southern Africa. In another example, Antina von Schnitzler examined the evidentiary practices and epistemologies

---

1 Others in this volume have examined how identities are made. With attention to the role of data in digital lending in contemporary Kenya, Park and Donovan (2022) argue that personhood may best be understood as "the result of particular ensembles of language, law, ethics, and infrastructures."

underlying a constitutional rights case demanding the right to water brought by a community of Soweto residents (von Schnitzler 2013, 2014, 2016). She presents an excellent example of technopolitics showing how technoscience and law become entangled in "metrologies of dignity" (von Schnitzler 2014, 342–44). These two STS-inspired studies of the politics, law, science, and technology in postapartheid South Africa sit alongside other analyses of the social lives of the same laws undertaken within the field of sociolegal studies such as on the right to water (Dugard 2008).

2   Setting the Local Constitutional Scene

The right of access to information became part of the South African legal system as part of the postapartheid changes making the country's governance and introducing non-racial democracy. South Africa engaged in a constitution-making process with two formal constitutions. The first was an interim Constitution that came into effect as all South Africans were able to vote for their choice of national government in April 1994. The second was a final Constitution, negotiated under the democratic conditions confirmed by the first, and taking effect in 1997. One legally explicit mention of something like the right of access to information occurred as part of the Constitutional Principles contained in the 1994 Constitution. Another came with the simultaneous entrenchment of the right to access to information in the Bill of Rights. As mentioned above, the PAIA was then enacted in 2000. There is a small but dense mostly legal doctrinal literature around the mechanisms designed to legally implement and entrench the right of access to information (Currie and Klaaren 2002; Klaaren and Penfold 2002; Cachalia 2017; Klaaren 2018; Peekhaus 2014; Bosch 2006; van Heerden, Govindjee and Holness 2014).

A great degree of the justification and legitimacy of the right of access to information in the South African context derived from a particular vision of citizenship and democracy that was associated with this right at the time of its introduction. In the articulation of this imaginary termed "open democracy," citizens are able to use information, to which they have access, in order to hold government primarily to account for the delivery (or non-delivery) of public socioeconomic services (Jagwanth 2002; Calland and Tilley 2002). The focus on service provision went so far as to term the right to access to information as a leverage right. A less prominent theme was citizens' need for information and education in order to be able to know their own interests—a proposition associated with access to knowledge and right to know discourses. As noted above, there was only a faint discursive tradition on something like a right to information extant in South Africa prior to constitutional democracy—and moreover this tradition was focused on countering the secrecy of the apartheid state (Mathews 1978). While right to information advocates did reflect explicitly on the degree to which their discourse was globally inspired and driven, these reflections often did not acknowledge this historical gap (Calland 2009). Moreover, this articulation of the significance of the right of access to information as a leverage right towards socioeconomic rights also overlapped with a recognition of South Africa as a developing country (Peekhaus 2014, 571).

One might argue that the somewhat instrumentalist cast of this open democratic vision of citizenship results in part from translating such a right from the West at a particular historical moment. The construction of right to information discourse with such a tight linkage to politics focused on the achievement of socioeconomic rights took place at a particular time in both global and national political contexts. At least in the United States and perhaps in other Western countries that have embraced neoliberal politics since the early 1980s, the transparency doctrine was beginning to drift away from its initial focus on creating a better, more responsive state, and embody less of a social democratic vision (Pozen 2018). What remained was the focus on the state working with the private sector as an efficient instrument. Nationally, a fundamental symbolic and constitutive shift was underway towards socioeconomic rights that apartheid withheld for the non-white population (Langford et al. 2013).

The concept of freedom of information was embedded within a vision of open democracy in the register of multiparty electoral democracy, albeit with an active and vital civil society. This vision of citizenship associated with the right of access to information has several key drivers or presumptions embedded within it additional to the distinct narrow question of whether a public organisation will grant a citizen access to information that the citizen has requested. One is the presumed competence of citizens to ask for a record, to communicate the request. Another is the capability to understand the information to which they have access. A third was the arguably necessary intermediation of civil society organisations to process and transform the information into knowledge and then into political action.[2]

---

2 The first postapartheid generation of civil society organisations associated with the right of access to information was led by the Open Democracy Advice Centre and included the South African History Archive. Now deceased, that organisation saw its primary mission as assisting the state in institutionalising the right of access to information (Calland 2009). In the second generation of access to information NGOs, the current leading civil society organisation advocating for in its terms "the right to know"—the Right to Know Campaign—is interestingly evolved out of but beyond opposition to secrecy, and is more focused on questions of identity, the knowledgeable citizen, and culture (Mottiar and Lodge 2020).

FIGURE 5.1   Palace of Justice, Tshwane
PHOTOGRAPH: BY KYLE-PHILIP COULSON ON UNSPLASH.COM

Each of these presumed preconditions may be questioned. To focus on the technological angle of just one of the above, one might ask what is the value and the reality of a legal right of access to information where the exercise of the right is itself dependent on the ownership or the provision of a technology such as a mobile phone to have that technology access information (Klaaren 2002). This question is equally pertinent in an economic register when the device at hand depends upon the price of data to operate (Competition Commission 2019). In this sense, an access-to-information statute can be linked to a techno-logical device such as a cell phone—both are mechanisms for access to infor-mation (van Heerden, Govindjee, and Holness 2014, 32).

Even more prominent now, but still present at the creation of South Africa's constitutional democracy was a strand of the globally circulating openness discourse (as well as on the ground social practices or forms of organisa-tion) distinct from the open democratic one tightly associated with the right of access to information as described above. This separate strand—we may term it "open government"—sees access to information less in the register of representative democracy and more in the registers (ironically themselves contradictory) of expert-led engineered democracy often associated with neoliberalism and, to a lesser degree, participatory or bottom-up democracy (Pozen 2018). In its South African instantiation, this separate strand of the global discourse focuses on achieving growth within the national economy in the democratic era (Plantinga and Adams 2021). This discourse could be built upon some progressive policy advances achieved under apartheid as well as in the democratic era (Hassim 2005).

3   Constructing the Promotion of Access to Information Act (PAIA)

The right of access to information in the 1996 Constitution was one of several rights that were selected for (or subjected to) mandatory legislative enforcement. This meant that the South African Parliament had a finite period of time—three years—in which to implement the right through drafting a parliamentary statute to put the right into effect. Simultaneously with the drafting of the interim Constitution, some civil society organisations had already been advocating for what they termed "an Open Democracy Bill" (Currie and Klaaren 2002, 7). Upon entering office, the ANC government appointed a task team to develop policy. In 1995, the task team recommended comprehensive legislation be drafted, including whistleblowing protection, open meetings (government in the sunshine) regulation, a law on correction and protection of personal information, and an access to information law. The Department of Justice formed a drafting team of lawyers with some governmental and administrative experience which consulted widely within government and looked carefully at the growing number of statutes (usually called Freedom of Information Acts) adopted by jurisdictions around the world (Currie and Klaaren 2002, 7-11). There was little if any attention in this process to the impact of electronic information technology on access to information produced by public bodies.

Over the course of its drafting, the framework for the statute narrowed considerably. From the initial Open Democracy Bill, the official legislation evolved to take the form essentially of the last element, a law providing for a citizen to request information from government bodies and to have a presumptive right to such information, although the right could be overridden by a legitimate need for government confidentiality. The legislation contained numerous grounds of refusals, apparently to accommodate each and every potentially legitimate (usually in the sense of being globally common) ground of refusal. The law lost its label of open democracy in the process of drafting and was renamed the Promotion of Access to Information Act. The final product was globally seen as a gold standard of freedom of information laws.

The overlap of the subject matter of access to information with that of privacy proved significant in the drafting of the PAIA, both in process and in substance. Concerns over privacy showed up in two areas. In order to protect privacy (and other grounds of refusal), PAIA mandated extensive third-party notification procedures, requiring that affected persons be told about and have a chance to contest disclosures of information. It also contained one single section obligating each government department to exercise a reasonable duty to facilitate corrections to personal information held by the state. Conspicuously missing was any true substance of personal information protection (what is often termed "data protection"). That part of the purpose of the Open Democracy Bill had attracted fierce opposition from the marketing and advertising industry.

The technological mindset within which the PAIA was drafted tilted towards paper documents. While not out of step with then-extant freedom of information laws around the globe, as mentioned above, PAIA did not pay any significant degree of attention to electronic technology. For instance, the statutory definition of its key implementing concept, a record, did not expressly refer to electronic data. Instead, the clear conceptual underpinning for the term "record" was that of a paper document. Nonetheless, the human rights-oriented expert commentators on the PAIA argued that information held in the form of electronic data—and which could be translated into a record through a "routine" query of the relevant database—should be included (Currie and Klaaren 2002, 42).

I soon learned that there was substantial ambiguity in how to interpret the text of the PAIA.[3] This meant some potential to influence the application of the PAIA. For example, one basis on which to choose among different legal interpretations of several important statutory provisions of the PAIA is the goal to further the well-functioning of the archives and records management systems of the early postapartheid South African state. One significant but ambiguous PAIA provision is found in sub-section 14 (1)(d). Here, PAIA requires public bodies to compile in at least three official languages a manual on the functions of, and index of records held by that body. The sub-section requires the manual to contain "sufficient detail to facilitate a request for access to a record of the body, a description of the subjects on which the body holds records, and

---

3   Allan's observation (2009, 147) that the legislation is "largely unambiguous" needs to be appreciated contextually. She goes on to state "... however, there has been little consistency in the approach followed by public bodies, and disputes about what the legislation intended have resulted" (2009, 147–148). While the PAIA does demonstrate clarity, precision, and detail in its drafting, ambiguity nonetheless is present, at least due to either inherently ambiguous terms or paradoxically the statute's detail and precision, which can result in mind-numbing degrees of complexity. See e.g. Centre for Social Accountability v Secretary of Parliament and Others (298/2010) [2011] ZAECGHC 33; 2011 (5) SA 279 (ECG); [2011] 4 All SA 181 (ECG) (28 July 2011).

the categories of records held on each subject." The specification of the actual information required to be published in the manual about the records held by the public body is vague. Further, as my co-author and I noted in our commentary on PAIA: "[t]he section's title raises the expectation of requirements for an 'index of records', but this is not fulfilled in the section's text" (Currie and Klaaren 2002, 217). Taking advantage of this ambiguity, we proposed a particular model to give meaningful expression to the terms 'subject' and 'category' as used in the section, a model that we thought would facilitate effective records management.

The implementation of section 14(1)(d) also implies work of classification, done in this case at the level of specific public bodies in the South African state. While there may be much social and political work (including expert commentary) that goes into the making of classification schemes of subjects and categories contained in these PAIA classification manuals, once made, the schemes appear natural and technical.[4] There are hundreds of these manuals extant, from both private and public bodies. Entangled now with other expert knowledges, the specification of meaning of sub-section 14(1)(d) was in part translated from the knowledge of the archivists' profession through the text of our commentary into the legal interpretation of the PAIA.

The rollout of the right of access to information and its implementing legislation was contested and often conflictual (Allan 2009; Peekhaus 2014). The most detailed and comprehensive treatment of the technical matters such as the appropriate interpretation of section 14(1)(d) associated with its implementation is Kate Allan's "Applying PAIA: Legal, Political, and Contextual Issues," a chapter in her edited collection *Paper Wars*. In addition to chronicling several disputes in which public bodies employed legal tactics to avoid disclosure in response to SAHA's request for records, Allan explores the technicalities of what she terms two issues internal to PAIA—enforcement mechanisms and the interaction of the PAIA with other information legislation—and then turns to look at external factors: the destruction of records, record-keeping practices, and cultures of transparency or secrecy "inherent in" public bodies.

The enforcement mechanisms provided for in PAIA are extensive in theory but pose extensive barriers and limitations in practice. This is for several reasons: that independent regulators have failed to respond to complaints, that there is little to no independent regulatory intervention following the internal appeal process, and that, where a public body is taken to court, bodies often settle prior to a precedent-setting decision (Allan 2009, 168–69). The PAIA originally placed

---

4 Johanna Mugler provides a brief overview of some social science literature on complex classification systems (Mugler 2019, 11–12).

obligations on two government bodies to regulate and to assist in enforcing the Act: the South African Human Rights Commission (SAHRC) and the Public Protector. Both were reported to have failed. The SAHRC justified its limited interventions on the grounds that it lacked resources and on the grounds that it was given a weak enforcement power in the PAIA—essentially only a power of recommendation, not a binding enforcement power (Allan 2009, 170). The ineffectiveness of the internal appeal mechanisms led to a call (discussed below) from PAIA advocates for a binding enforcement power (Allan 2009, 173).

In 2004, the access to information regime built around the PAIA as its lodestar was "incomplete" and yet as late as 2007 civil society figures committed to access to information retained hope. By 2010, the writing was on the wall, even to its advocates: the PAIA was charitably assessed as a law still in its starting blocks. In 2018, there was "no significant improvement." In 2020 one local researcher of public administration was referring to its "notorious" non-compliance among public bodies as an accepted fact (Klaaren 2006, 167; Caldwell 2010; Zulu 2018, 4; Pearson 2020; Calland 2009).

Access to information advocates endured the slow and painful realisation that the extensive legalism, clunky regulatory mechanisms,[5] and protracted timelines of access to information requests in terms of the PAIA had resulted in a law that was not only difficult to implement but actually arguably counter-productive to the Act's professed purpose of facilitating access to information (Cloete and Auriacombe 2008). This counter-productive effect happens in both formal and informal ways—for example, in the sense of distracting and using up needed and scarce organisational resources within public bodies and in the sense of providing an excuse not to disclose government information. The protection of privacy as implemented in the PAIA and interpreted by public bodies has also contributed to PAIA's implementation difficulty (Allan 2009, 150–57). For instance, the Department of Justice persisted in using the PAIA third-party notification procedures (which are designed to protect privacy of third-parties in records held by the state by allowing those parties to either consent to the release of the records or opposed their release) in situations,

---

5   One episode of particularly misguided bureaucratic implementation saw thousands of hard copies of PAIA manuals initially pile up in a storeroom at the SAHRC. This was due to a PAIA provision mandating such a move even for small businesses and individuals working as independent contractors, professionals, or consultants and the failure to decree an exemption. An exemption was later promulgated and remained in place for ten years until the provision was eliminated by amendment in the process of establishing the Information Regulator. See Adams and Adeleke 2020.

as Allan explains, where the requested records "have been aired in public hearings or are in the public domain and privacy rights have therefore lapsed" (Allan 2009, 150).

4   The PAIA Meets Three Information Technologies

This section narrates two periods that can be marked by the relative prominence of three separate information technologies within the South African public administration: paper documents, electronic databases, and digital ecosystems. The term "digital ecosystem" or "ecosystem" is used here as it is used in the information technology field. That limited and particular meaning refers to digitally interconnected sets of services through which users engage in a variety of uses in one integrated experience. The period from the end of apartheid to say 2008 may be thought of as revealing a shift in the imagined legal purpose of the right of access to information resulting in a technological shift from paper documents to large electronic databases. Then, the period from that point in time to the present (e.g. from 2008 to 2020) can be thought of as revealing a similar shift from databases to ecosystems. The two periods also are marked by a change in the specification of an enforcement agency for the right of access to information external to the public administration. At PAIA's enactment, the first line enforcement body was the semi-constitutional body, the South African Human Rights Commission. In 2013, PAIA was amended by POPIA to have as its first line enforcement body a statutory body, the Information Regulator.

4.1   PAIA and its Records from 1994 to 2008

Once enacted and placed on the statute book in 2000, the PAIA became part of South African public administrative law. While it was now legally enforceable, albeit only in the courts, it had also lost some of its constitutional status. PAIA was merely one of a number of newly drafted public administration-wide statutes that bureaucrats working at state organisations would need to fulfil and comply with (Mugler 2019, 8). Even for PAIA advocates, it was understandable that individual officials deciding on requests for records would at times fail to comply with the PAIA, and instead comply with either the public service regulations disciplining officials for disclosing confidential information, or with still legally enforceable secrecy laws (Calland 2009, 7).

How did officials of the South African public administration understand the meaning of PAIA and the democratic ideals behind its passage as they complied (or not) with its provisions? Let us start at some distance from the PAIA with an ethnographic study of one particular postapartheid state organisation, the National Prosecuting Authority (NPA). In an anthropological study, Johanna Mugler has examined how another postapartheid piece of public administrative law, the Public Finance Management Act, affected the accountability practices and internal self-identity of the NPA (Mugler 2019). Mugler's ethnographic research is able to link the stories told about accountability (for instance about when a prosecution should go forward and when it should be deferred) by the NPA's employees to the more official stories told by the NPA itself through the mechanism of the Public Finance Management Act (PFMA). Mugler pays particular attention to the numbers generated by the performance management systems mandated by the PFMA, exploring those numbers against the background of an academic literature concerned with the power of numbers and indicators (Rottenburg et al. 2015; Mugler 2019). One of Mugler's findings is that by ten years after the passage of the PFMA (around 2008), the "performance statistics, indicators, rates and targets form[ed] a central part of prosecutors' understanding and practices of accountability" (Mugler 2019, 90).

It would be impossible to do an ethnographic study (such as Mugler did) on the whole of the South African public service. However, we can use the interpretation and implementation of three particular sections (14, 15, and 32) of the PAIA to substitute for Mugler's monograph and thus to construct a rough portrayal of the co-production of the entire PAIA within the entire public service. In the interpretation and implementation of each of these three provisions, specific organisational choices—such as allocating different amounts to unit budgets, changing organisation procedures, or altering staff job-descriptions—were necessary to take (or to avoid) in creating the PAIA practice within specific public bodies.

This section has investigated in some detail the first of these provisions in the section above. Section 14's mandate for a manual with a classification of the subjects and categories of records was one location where technical experts could choose to align their production of technical knowledge with the open democratic imaginary of PAIA's founding. The remainder of this section covers the other two particular PAIA sections illustrative of its co-production: section 32 and section 15.

Second, while it never achieved the bureaucratic power of the PFMA nor did it depend as centrally on the power of numbers (and those two points may well be interrelated), the PAIA presented itself (like the PFMA both inside and outside the National Prosecuting Authority) both as a law which needed to be complied with and also as a law providing at least one tool for demonstrating and measuring such compliance. This tool was a mechanism provided for in section 32, entitled *Reports to Human Rights Commission*. This text required the information officer of each public body to submit numbers annually to the Human Rights Commission reporting on activities such as the number of requests received, granted, and refused as well as internal appeals lodged and applications to court made. Mandating a report that would be delivered external to the public body, section 32 gave the responsible public officials an opportunity to articulate an understanding of the PAIA from the perspective of their organisation. Moreover, providing for a reporting mechanism to measure

FIGURE 5.2   Church Square Precinct in Pretoria, South Africa. Pretoria has been the national administrative capital from 1910, with nearly all national department headquartered in the city, now renamed Tshwane.
PHOTOGRAPH: BY SIPHO NDEBELE ON UNSPLASH.COM

performance potentially allowed for some coordination of public body compliance as well as oversight of the potentially contradictory ways in which organs of state might comply. To some extent, the measurement and metrics of section 32 have been replicated in the Shadow Report practice engaged in by a network of South African access to information NGOs (Zulu 2018).

While the PAIA's number-driven accounting could conceivably have become as powerful as PFMA's in the operations of the NPA as Mugler documents, that is simply not what happened, at least not across the great bulk of the South African public administration. In the SAHRC's annual report on this section 32 reporting, the majority of public bodies appear not to have provided any information at all to the Commission. Albeit maintaining the numbers as a substratum, the SAHRC has resorted to representing the lack of compliance with the PAIA through colour coding the figures in the tables provided in its annual reporting. Thus, the SAHRC now reports green for any number zero or greater and red for no number reported. In an absurd touch, zero was coloured green as a good sign that at least the public body satisfied section 32 of PAIA and reported a number. In the SAHRC's latest (and last) report, around half the table is red (South African Human Rights Commission 2020). The Report noted that "[o]nly 20 out of 46 national departments submitted section 32 reports to the Commission. The level of compliance at the national level is at its lowest since 2010/11" (South African Human Rights Commission 2020, 27).

A third illustrative PAIA provision is section 15, covering the proactive provision of information such as through a website or other technological means. Despite PAIA's near complete lack of explicit engagement with electronic data, PAIA practitioners had some small hope and initial expectation that electronic technology would assist access to information. PAIA advocates were able to point to PAIA section 15. According to section 15, proactively disclosed information (such as information made available by a public body on its website) would not be subject to any request for access in terms of the Act, since it would already be in the public domain. PAIA contained very few rules mandating publication of information. Instead, section 15 was thought to provide an incentive for government departments to proactively publish public information under their control. The incentive was that since such information would not be subject to the PAIA request regime, it would be more efficient for public bodies to publish such information rather than hold it and spend resources complying with a PAIA request. The expectation behind section 15 also aligned with the expectation that government departments would increasingly grow to proactively rather than reactively disclose information. Indeed, this aspect of compliance was included in the mandatory annual reporting to the Human Rights Commission.

Some government units did pursue this section 15 route of information disclosure with vigour, pushing openness and using technology to do so. As one might expect, the Constitutional Court and its website was a shining and much-held-up example, particularly in the late 1990s and early 2000s (Constitutional Court of South Africa 2022). Much if not most of the publicly available information from the Constitutional Court is available on its website for the public to access and to access without needing to make a specific PAIA request. Likewise, organisations in the private sector or at least with one foot in the private sector such as universities have also implemented section 15. For instance, the University of the Witwatersrand identifies significant categories of information that are available in a category of "voluntary disclosure" such as records publicly available on the university's websites, as well as other categories of records that are available without a PAIA request (such as requests for personal information) (University of the Witwatersrand 2021). But most public bodies, such as the Department of Home Affairs, have not elaborated upon the voluntary disclosure model of section 15 and merely list on a single page the headings of the official department webpage (Department of Home Affairs 2013).

4.2   *The Change from Records to Databases*

As pointed out above, the practice of access to personal information was hobbled from its start, carving out from the Open Democracy Bill in the making of the PAIA any legal basis for the protection of personal information. The practice of obtaining access to and protection of one's personal information has nonetheless grown. Much of this is more than likely due to the regulation of data protection inhering in private corporate practices as well as the default settings and mechanisms of ICT devices. Some of this can nonetheless be ascribed to the moral force of the fight against apartheid as numerous persons requested their previously secret personal information retained in the files of the erstwhile state security. From the 1990s, the primary civil society organisation pursuing this work has been the South African History Archive (SAHA). Ironically, SAHA continually ran into the privacy ground of refusal (PAIA section 34) when submitting its requests for the files of former anti-apartheid activists and others surveilled by apartheid security agents (Allan 2009, 150). Rather than battling a discourse of confidentiality and a culture of privacy, SAHA's chief substantive opponent was apartheid's tradition of secrecy. Verne Harris characterised this as late as 2007 as: "... cultures of secrecy are proving extremely resilient. These ... cultures do not flow only out of the old apartheid state milieus. They also flow out of the anti-apartheid experiences of exile, the underground, and mass resistance" (Harris 2009, 211).

Harris was speaking about access to personal information from the viewpoint of open democracy. But there is a different and equally valid way to pose the relationship of notions of secrecy, confidentiality, opacity, and privacy to those of openness, transparency, disclosure, and access to information. Instead of constituting the exception, secrecy, and its related norms, may be seen as the rule.

By around 2008, the discourse surrounding privacy and the access to personal information had become dominant in constitutional democratic debate over openness and secrecy. No longer merely one part of an overarching access to information framework, privacy rights advocates began to push a data protection framework that would come to see access to information as an exception to its rule. The need for a Data Protection Act began to be again acknowledged as early as 2005 in public discussions with government.[6]

The growing perception of the need for data protection stemmed from both legal and technological locations. First, the implementation of the access to information law was going badly. Opting for a legal drafting solution, civil society organisations articulated the need for a statutory body more accessible than a court to make a binding enforcement order (Allan and Currie 2007). The proposed protection of personal information legislation became the vehicle for achieving this binding independent enforcement power demanded and called for by right to access to information advocates (Allan 2009, 173).

Second, within the public administration, the dominant information technology had shifted from the use of documents to the use of electronic databases. The idea was now to run the public administration on the basis of large capacious and effective databases. Writing in 2005, Breckenridge observed that "... the [South African] state's interest in digital biometrics is very largely driven by a desire to repair a broken bureaucracy, to deliver grants and other benefits to the poorest and most vulnerable of its citizens" (Breckenridge 2005, 270). Indeed, in the middle of the first decade of the twenty-first century, South Africa was doing relatively well in developing large-scale databases in both the public and the private spheres, embarking upon several very large projects in the sectors of transport, justice, and social protection. Digitisation held the realistic prospect of facilitating the achievement of socioeconomic rights in South Africa. While the very notion of a democratic state comes with the necessity to know the population, on the other side of this coin was the danger of overreaching intrusion upon individual privacy. These large electronic

---

6 See Park and Donovan (this volume) for a discussion of similar data protection legislation in Kenya.

database projects played a significant contributing role in the emergence of privacy protection as a dominant discourse in the mid-2000s.

Third, the articulated need for privacy legislation specifically in the form of a Data Protection Act was becoming stronger at domestic as well as global levels. Even PAIA advocates became convinced of the principled need for privacy legislation in this form (Calland 2009). The new element that put this legislation on the political agenda was the desire for legislation to be in place to protect the privacy rights of data subjects contained in databases of information moving across borders. The theme of data privacy was emerging as a strong player in the dynamics of openness and secrecy. Whereas the demand of the marketing and advertising industry in the 1990s had been against data protection legislation, the balance of forces within that media sector had changed by the first decade of the twenty-first century and such a law was now seen as a necessary part of a modern industrial nation.

Shifting from open democracy to open government, the domestic privacy discourse was changing in the direction of more globally mainstream concerns for data protection rights. This was manifested in institutional form as the privacy act drafting process unfolded. As Rachel Adams and Fola Adeleke (2020) argue, the parliamentary decision to opt for an economic regulatory body (Information Regulator) rather than a semi-constitutional rights enforcement body (the Human Rights Commission) was a significant retreat from the initial aspirations of the democratic constitutional regime. Revealingly, a report of the South African Law Reform Commission on privacy legislation pegged the need for such data protection legislation to market forces (Adams and Adeleke 2020).

4.3   *PAIA from 2008 to 2021*

The story over PAIA's years from 2008 to 2021 varied sharply from its initial period. In 2008, a drafting process began that would result in the passage of data protection legislation, the POPIA.[7] The law was formally enacted in 2013 and regulated data protection and privacy as well as setting up an Information Regulator with responsibility over both access to information and data protection/privacy. The Regulator herself was appointed in 2016 and the Office of the Information Regulator began to be resourced and to take shape around her from that point (Adams and Adeleke 2020). On July 1, 2021, the Information Regulator took over regulatory functions relating to PAIA from the South African Human Rights Commission (Polity 2021). Twenty-five years after the

---

7   The legal argument that such legislation should have been drafted alongside the PAIA was noted but passed over in the SALRC report developing data protection legislation.

interim Constitution, a non-court enforcement mechanism for access to information was finally albeit formally established.

During this second period of PAIA's construction, there were further changes within the realm of information technology, both globally and within the South African public administration. Globally, the focus of information technology shifted away from large-scale electronic databases to the integration of information technologies, cloud computing and the creation of digital ecosystems. The end point of this trend can be seen for instance in key areas of the public administration. By 2021, a donor organisation partnering with the National Treasury was advertising a position for an "embedded technical consultant" as part of reforming the field of public procurement, a government practice running just under half of the state budget. As the advert states, the idea is for the hire to participate in "... playing a pivotal role through an eco-system approach to support the National Treasury to make procurement more transparent, effective, and resilient as a key part of South Africa's economic recovery efforts to address the corruption risks inherent with procurement" (Open Contracting Partnership 2021).

The technological shift can be seen in the content of the various industry standards. In the ICT sector, industry standards (ratified after the fact by public bodies) rather than rules produced and disseminated by public bodies themselves are an increasingly significant mode of regulation and governance internationally (Baron and Spulber 2018; Bennett and Raab 2020). Both in public and in private international circles, digital technology policy developed during the first two decades of the twenty-first century in ways that were largely deaf to voices from the developing countries (Kira 2020).

South African regulation of ICT has followed (at a distance) rather than kept up with these global developments in information technology governance. South Africa's e-government strategy has not been updated since 2001 and the two key sets of minimum standards on information security and interoperability have not been updated since 1996 and 2008 respectively. Non-compliance by government agencies with even the existing standards is significant. It is thus likely that what compliance to standards is, occurring is to standards comprising part of transnational private regulation (Manda and Backhouse 2016). For instance, South Africa has not been able to implement a key technology policy adopted in 2003 on free and open source software. These software systems were hardly adopted within government as departments continued to procure commercial off-the-shelf products (Ngoepe 2015). The primary reason for the lack of free and open source software adoption reported by government users was incompatibility between such software and existing proprietary technologies (Mtsweni and Biermann 2008).

Arguably, this failure to govern technology at this level of industry standards is a failure of South African democracy. The problems of encouraging openness and of enabling the coordination of public information infrastructures were to be addressed by the State Information Technology Agency (SITA), established in 1999. Even after the amendment of its empowering legislation to adopt standards as a mode of regulation, SITA's legislation still largely views regulation as a mode of ministerial command and control, within a public service paradigm. First promulgated only in 2008, SITA's Minimum Interoperability Standards (MIOS) do not cater for coordination among government departments and are primarily used in and directed at hardware procurement (State Information Technology Agency 2021).

The democratic aspirations carried by the POPIA are significantly reduced compared to those of the right of access to information. The mission to protect personal information is hardly the same as that of providing access to information. The Information Regulator began to enforce the POPIA from July 1, 2021. While it has a variety of regulatory mechanisms, the POPIA arguably reflects its time of initial drafting and is focused more on databases than on information ecosystems. And effective regulation of informational ecosystems arguably requires collaboration between (at least) data protection and the competition authorities (to deal with the corporate dominance of big tech). In South Africa, such collaboration has appeared on the regulatory agenda but effective institutions remain not yet built (Competition Commission South Africa 2021, 2020).

The institutional position of the Information Regulator as a statutory regulatory body focused on the economy rather than as a semi-constitutional body supporting democracy fits hand in glove with a significant discursive shift that has recently become apparent. The ideal of a functioning representative democracy at the heart of the open democratic vision of citizenship and focused on the achievement of socioeconomic rights through the mechanisms of an effective state has waned (Sibanda 2011; Madlingozi 2017). Open government globally has come to be closely associated with innovation for inclusive development (Plantinga and Adams 2021). This new doctrine of open government envisions mechanisms aimed at achieving inclusive development outcomes including: open interaction between government, technologists, and end-users in the design of policies and services; open, market-oriented approaches to delivering public services and meeting the needs of citizens; and a central role for information and communication technologies (ICTs) in policy development and innovation activities. The civic tech movement of the current digitisation moment appears to have a focus on electronic government rather than on information government (Mayer-Schönberger and Lazer 2007). The questions should be asked whether hybrid civic-technology innovation networks at the fringes of formal open government initiatives can sustain meaningful interaction with mainstream political processes and how public officials may engage with such movements in order to meet national development outcomes (Plantinga and Adams 2021).

There are democratic risks and dangers apparent in this development. The shiny promises of open government as hardware/software may distract from the social capital aspects (skills and training) of the state's information infrastructure. Digitalisation may be used as an intellectual crutch by state officials to avoid the maintenance work necessary for effective public service, as when records management professionals superficially perceive digital information technology as a panacea (Katuu 2015). Moreover, the adoption of open, ICT-oriented practices risks shifting responsibility and trust for governance and development outcomes, from a somewhat accessible national democratic state, to largely unaccountable global and national network enterprises. Arbitraging among legal frameworks as well as information systems, well-resourced private corporations may be best able to take advantage of this environment, contributing to global and domestic inequalities.

Similar to other comparable countries such as Brazil or India, the contemporary public information practice in South Africa is diverse and fragmented, consisting of documents, databases, and ecosystems. Whatever the cause of this state of affairs,[8] it seems important to recognise that this unevenness of South African public administration indicates that these three information technologies have settled or moved into a scheme where they operate alongside each other. Certainly, paper documents still play an important and at times dominant role. For instance, according to one well-researched and vividly written recent article, "documents form ... central instruments in advancing competing narratives in state institutions, potential instruments in strategies of 'Orwellian manipulation'" (Pearson 2020, 79). However, this author carefully noted that "most of the documents cited ... were not readily accessible, and in fact required careful and extensive negotiations between PARI researchers and the municipal administration. Large swathes of official documents which are central to how power is exercised and contested in state institutions remain locked away." While PAIA may have been turned into a dead letter in most of the units of the public administration, that does not mean it has been entirely forgotten, even in this later period. For this author, "opening state archives to

---

8   Cloete (2012) has argued that South Africa faced constraints on its development of e-government that other developing countries did not. These constraints include "a lack of political will and support; a lack of strong and consistent leadership; a weak and contradictory IT governance framework; and continuous political and bureaucratic infighting" (Cloete 2012, 138).

public scrutiny constitutes one important step towards ensuring that the 'complex and ever-shifting power plays' that unfold among state functionaries are balanced by democratic impulses."

5   Conclusion

Recognising and demonstrating the dialectics between the goals of openness and secrecy, this chapter has elaborated how legal attempts to pursue the rights of access to information and privacy have co-produced a sociopolitical order together with three information technologies. That sociopolitical order is the PAIA. Obviously, PAIA is impossible to understand without talking about it using legal terms. In South Africa's initial transition from apartheid to constitutional democracy, a discourse of open democracy justified the constitutional right to information and its associated vision of citizenship including the achievement of socioeconomic rights. Required by the Constitution, a formal law was quickly put onto the statute books and served as the location for much legal work of drafting, interpretation, and, to a lesser degree, enforcement. The closely associated right of privacy complicated but did not derail this legal work. Largely ignored by public officials from the start, PAIA's enforcement powers were later amended and strengthened through the enactment of the Protection of Personal Information Act in 2013. This was a demonstration of the increasing prominence and power of data protection in global and domestic discourses of transparency and privacy.

This chapter has argued that it is equally impossible to comprehend the full reality of the PAIA and understand how it became what it is without viewing it from the perspective of science and technology studies. PAIA's construction, meaning, and power are not solely formal legal matters but are deeply entangled with technical knowledge and with different forms of information technology. The (limited) extent of PAIA's power and reach over South Africa's bureaucracy cannot be understood without attending to the expert technical knowledge contributing to its meaning, without thinking through the exercise of classification and categorisation of information that it necessitates, and without looking closely at the instruments of measurement it sets up to remain accountable (at least as a formal matter) within the South African state. PAIA is further in great degree a production of three information technologies used in various degrees and various bodies of the South African public administration. Intended to serve as a powerful tool for access to paper documents, PAIA soon encountered the use of two further state information technologies: large electronic databases and, most recently, digital information ecosystems.

The co-production of the current PAIA social order can be best seen as occurring over two periods of construction: the first from constitutional democracy to the prominent role of large electronic databases in the South African public administration around 2005 and the second from that point to the present. At present, public administrative bodies are increasingly employing digital information ecosystems alongside paper documents and alongside electronic databases. As illustrated by the PAIA, the ideals of open democracy have waned and the linked discourses of open government and data protection are ascendant. Questions about the shape, character, and sustainability of the citizenship and democracy associated with these developments are worth posing. In line with the argument presented here, the answers are likely to be co-produced by law and information technology.

Bibliography

Adams, Rachel and Fola Adeleke. 2020. "Protecting Information Rights in South Africa: The Strategic Oversight Roles of the South African Human Rights Commission and the Information Regulator." *International Data Privacy Law* 10, no. 2: 146–59.

Allan, Kate. 2009. *Paper Wars: Access to Information in South Africa*. Johannesburg: Wits University Press.

Allan, Kate and Iain Currie. 2007. "Enforcing Access to Information and Privacy Rights: Evaluating Proposals for an Information Protection Regulator for South Africa: Current Developments." *South African Journal on Human Rights* 23, no. 3: 570–86.

Baron, Justus and Daniel F. Spulber. 2018. "Technology Standards and Standard Setting Organizations: Introduction to the Searle Center Database." *Journal of Economics & Management Strategy* 27, no. 3: 462–503.

Bennett, Colin J. and Charles D. Raab. 2020. "Revisiting the Governance of Privacy: Contemporary Policy Instruments in Global Perspective." *Regulation & Governance* 14, no. 3: 447–64.

Birchall, Clare. 2011. "Transparency, Interrupted Secrets of the Left." *Theory, Culture & Society* 28, no. 7–8: 60–84.

Bosch, Shannon. 2006. "IDASA v ANC – An Opportunity Lost for Truly Promoting Access to Information." *South African Law Journal* 123, no. 4: 615–25.

Bozzoli, Geurino Renzo 1975. "Academic Freedom in South Africa." *Minerva* 13, no. 3: 428–65.

Breckenridge, Keith. 2005. "The Biometric State: The Promise and Peril of Digital Government in the New South Africa." *Journal of Southern African Studies* 31, no. 2: 267–82.

Brunette, Ryan. 2014. "The Contract State: Outsourcing & Decentralisation in Contemporary South Africa." Johannesburg: Public Affairs Research Institute. http://www .pari.org.za/wp-content/uploads/PARI-The-Contract-State-01082014.pdf.

Cachalia, Firoz and Jonathan Klaaren. 2022. "Digitalisation in the Health Sector: A South African Public Law Perspective." *Potchefstroom Electronic Law Journal* 25: 1–24

Cachalia, Raisa. 2017. "Botching Procedure, Avoiding Substance: A Critique of the Majority Judgment in My Vote Counts." *South African Journal on Human Rights* 33, no. 1: 138–53.

Calavita, Kitty. 2010. *Invitation to Law and Society: An Introduction to the Study of Real Law*. Chicago: University of Chicago Press.

Caldwell, Marc. 2010. "Information Wars: Kate Allen, The Paper Wars: Access to Information in South Africa." *Journal of Southern African Studies* 36, no. 2: 511–13.

Calland, Richard. 2009. "Illuminating the Politics and the Practice of Access to Information in South Africa." In *Paper Wars: Access to Information in South Africa*, edited by Kate Allan, 1–16. Johannesburg: Wits University Press.

Calland, Richard and Alison Tilley, eds. 2002. *The Right to Know, the Right to Live: Access to Information and Socio-Economic Justice*. Cape Town: ODAC.

Cloete, Fanie and Christelle Auriacombe. 2008. "Counter-Productive Impact of Freedom of Access to Information-Related Legislation on Good Governance Outcomes in South Africa." *Journal of South African Law / Tydskrif Vir Die Suid-Afrikaanse Reg*, no. 3: 449–63.

Competition Commission South Africa. 2019. "Data Services Market Inquiry: Final Report: Summary of Findings and Recommendations." http://www.compcom.co.za/wp-content/uploads/2019/12/Data-Market-Inquiry-SUMMARY.pdf.

Competition Commission South Africa. 2020. "Competition in the Digital Economy (Version 2)." http://www.compcom.co.za/wp-content/uploads/2021/03/Digital-Markets-Paper-2021-002-1.pdf.

Competition Commission South Africa. 2021. "Online Intermediation Platforms Market Inquiry Terms of Reference." https://www.compcom.co.za/wp-content/uploads/2021/04/44432_09-04_EconomicDevDepartment.pdf.

Constitutional Court of South Africa. n.d. "Home: Constitutional Court of South Africa." Accessed October 30, 2022. https://www.concourt.org.za/.

Currie, Iain and Jonathan Klaaren. 2002. *The Promotion of Access to Information Act Commentary*. Cape Town: Siber Ink.

Department of Home Affairs. 2013. "Department of Home Affairs – PAIA MANUAL."

Dugard, Jackie. 2008. "Rights, Regulation and Resistance: The Phiri Water Campaign." *South African Journal on Human Rights* 24, no. 3: 593–611.

Foster, Laura A. 2017. *Reinventing Hoodia: Peoples, Plants, and Patents in South Africa*. Seattle: University of Washington Press.

Harris, Verne. 2009. "Conclusion: From Gatekeeping to Hospitality." In *Paper Wars: Access to Information in South Africa*, edited by Kate Allan, 201–13. Johannesburg: Wits University Press.

Hassim, Shireen. 2005. "Turning Gender Rights into Entitlements: Women and Welfare Provision in Postapartheid South Africa." *Social Research* 72, no. 3: 621–46.

Heerden, A van, A Govindjee, and D Holness. 2014. "The Constitutionality of Statutory Limitation to the Right of Access to Information Held by the State in South Africa." *Speculum Juris* 28, no. 1: 27–54.

Horn, Eva. 2011. "Logics of Political Secrecy." *Theory, Culture & Society* 28, no. 7–8: 103–22.

Jagwanth, Saras. 2002. "The Right to Information as a Leverage Right." In *The Right to Know, the Right to Live: Access to Information and Socio-Economic Justice*, edited by Richard Calland and Alison Tilley. Cape Town: ODAC.

Jasanoff, Sheila. 2004. *States of Knowledge: The Co-Production of Science and the Social Order.* London: Routledge.

Jasanoff, Sheila. 2009. *Science at the Bar: Law, Science, and Technology in America.* Cambridge, MA: Harvard University Press.

Katuu, Shadrack. 2015. "Managing Records in South Africa's Public Sector – A Review of Literature." *Journal of the South African Society of Archivists* 48: 1–13.

Kira, Beatriz. 2020. "Governing a Globalised Digital Economy: How to Make Technology Policy and Regulation Work for Developing Countries." *Global Policy* [online], August. https://papers.ssrn.com/abstract=3686777.

Klaaren, Jonathan. 2002. "A Right to a Cellphone?: The Rightness of Access to Information." In *The Right to Know, the Right to Live: Access to Information and Socio-Economic Justice*, edited by Richard Calland and Alison Tilley, 18–26. Cape Town: ODAC.

Klaaren, Jonathan. 2006. "The Right of Access to Information at Age Ten." In *Reflections on Democracy and Human Rights: A Decade of the South African Constitution*, 167–71. Johannesburg: South African Human Rights Commission.

Klaaren, Jonathan. 2015. "The South African 'Secrecy Act': Democracy Put to the Test." *Verfassung Und Recht in Übersee / Law and Politics in Africa, Asia and Latin America* 48, no. 3: 284–303.

Klaaren, Jonathan. 2018. "My Vote Counts and the Transparency of Political Party Funding in South Africa." *Law, Democracy, & Development* 22, no. 1: 1–11.

Klaaren, Jonathan. 2021. "Legal Aid SA: A Successful Post-Apartheid Institution Supporting the Rule of Law." In *Making Institutions Work in South Africa*, edited by Daniel Plaatjies, 171–87. Pretoria: Human Sciences Research Council.

Klaaren, Jonathan and Glenn Penfold. 2002. "Access to Information." In *Constitutional Law of South Africa*, edited by Stu Woolman, Second Revised and Enlarged Edition, 1–24. Kenwyn, South Africa: Juta Legal and Academic Publishers. https://constitutionallawofsouthafrica.co.za/wp-content/uploads/2018/10/Chap62.pdf.

Langford, Malcolm, Ben Cousins, Jackie Dugard, and Tshepo Madlingozi. 2013. *Socio-Economic Rights in South Africa: Symbols or Substance?* Cambridge, MA: Cambridge University Press.

Madlingozi, Tshepo. 2017. "Social Justice in a Time of Neo-Apartheid Constitutionalism : Critiquing the Anti-Black Economy of Recognition, Incorporation and Distribution." *Stellenbosch Law Review* 28, no. 1: 123–47.

Manda, More Ickson and Judy Backhouse. 2016. "Addressing Trust, Security and Privacy Concerns in e-Government Integration, Interoperability and Information Sharing through Policy: A Case of South Africa." *Proceedings of the International Conference on Information Resources Management (CONF-IRM)*.

Mathews, Anthony S. 1978. *The Darker Reaches of Government: Access to Information about Public Administration in Three Societies*. Cape Town: Juta.

Mayer-Schönberger, Viktor and David Lazer. 2007. *Governance and Information Technology: From Electronic Government to Information Government*. Cambridge, MA: MIT Press.

McCann, Michael W. 1994. *Rights at Work: Pay Equity Reform and the Politics of Legal Mobilization*. Chicago: University of Chicago Press.

Mottiar, Shauna and Tom Lodge. 2020. "'Living Inside the Movement': The Right2Know Campaign, South Africa." *Transformation: Critical Perspectives on Southern Africa* 102, no. 1: 95–120.

Mpofu-Walsh, Sizwe. 2021. *The New Apartheid*. Cape Town: Tafelberg.

Mtsweni, Jabu and Elmarie Biermann. 2008. "An Investigation into the Implementation of Open Source Software within the SA Government: An Emerging Expansion Model." In *Proceedings of the 2008 Annual Research Conference of the South African Institute of Computer Scientists and Information Technologists on IT Research in Developing Countries: Riding the Wave of Technology*, 148–58. SAICSIT '08. New York: Association for Computing Machinery.

Mugler, Johanna. 2019. *Measuring Justice: Quantitative Accountability and the National Prosecuting Authority in South Africa*. Cambridge: Cambridge University Press.

Naidoo, Vinothan. 2019. "Transitional Politics and Machinery of Government Change in South Africa." *Journal of Southern African Studies* 45, no. 3: 575–95.

Ngoepe, Mpho. 2015. "Deployment of Open Source Electronic Content Management Software in National Government Departments in South Africa." *Journal of Science & Technology Policy Management* 6, no. 3: 190–205.

Nuttall, Sarah, and Achille Mbembe. 2015. "Secrecy's Softwares." *Current Anthropology* 56, Supplement 12: S317–24.

Open Contracting Partnership. 2021. "Embedded Technical Consultant: National Treasury South Africa at Open Contracting Partnership." https://open-contracting -partnership.breezy.hr/p/643f9ddcab84-embedded-technical-consultant-national -treasury-south-africa.

Park, Emma and Kevin P. Donovan. Forthcoming. "Privacy, Privation and Person: Data, Debt and Infrastructured Personhood."

Pearson, Joel. 2020. "Paperwork and Power Plays: Contestation and Performance at a Limpopo Municipality." *Transformation: Critical Perspectives on Southern Africa* 103: 59–82.

Peekhaus, Wilhelm. 2014. "South Africa's Promotion of Access to Information Act: An Analysis of Relevant Jurisprudence." *Journal of Information Policy* 4: 570–96.

Plantinga, Paul, Rachel Adams and Saahier Parker. 2019. "AI Technologies for Responsive Local Government in South Africa." In *Global Information Society Watch 2019: Artificial Intelligence: Human Rights, Social justice and Development*. New York: Association for Progressive Communications: 215–220.

Plantinga, Paul and Rachel Adams. 2021. "Rethinking Open Government as Innovation for Inclusive Development: Open Access, Data and ICT in South Africa." *African Journal of Science, Technology, Innovation and Development* 13, no. 3: 315–23.

Polity. 2021. "Information Regulator Takes over PAIA Functions." Last modified June 30, 2021. https://www.polity.org.za/article/information-regulator-takes-over-paia -functions-2021-06-30.

Pozen, David E. 2018. "Transparency's Ideological Drift." *Yale Law Journal* 128, no. 1: 100–65.

Ravigopal, Bhavya. 2019. "The Impact of Intellectual Property and Regulatory Law on the Propagation of Medical Products to Various Countries." *Intersect: The Stanford Journal of Science, Technology, and Society* 13, no. 1.

Robins, Steven L. 2008. *From Revolution to Rights in South Africa: Social Movements, NGO s & Popular Politics After Apartheid*. Boydell & Brewer Ltd.

Rosenberg, Gerald N. 2008. *The Hollow Hope: Can Courts Bring About Social Change? Second Edition*. Chicago: University of Chicago Press.

Rottenburg, Richard, Sally E. Merry, Sung-Joon Park, and Johanna Mugler, eds. 2015. *The World of Indicators*. Cambridge: Cambridge University Press.

Sibanda, Sanele. 2011. "Not Purpose-Made! Transformative Constitutionalism, Post-Independence Constitutionalism and the Struggle to Eradicate Poverty." *Stellenbosch Law Review* 22, no. 3: 482–500.

South African Human Rights Commission. 2020. "SAHRC PAIA Report 2019–2020." https://www.sahrc.org.za/home/21/files/SAHRC%20PAIA%20Report%202019_20 _Final.pdf.

State Information Technology Agency. n.d. "Minimum Interoperability Standards Revision Version 5.0." Accessed October 30, 2022. http://rfq.sita.co.za/standard/Search able%20MIOS%20V5%20-%20Unsigned.pdf.

University of the Witwatersrand, Johannesburg. 2021. "Promotion of Access to Information Act User Manual, University of the Witwatersrand, Johannesburg". https://www.wits.ac.za/media/wits-university/footer/about-wits/paia/documents /Section_14_Manual_June%202021.pdf.

Von Schnitzler, Antina. 2013. "Traveling Technologies: Infrastructure, Ethical Regimes, and the Materiality of Politics in South Africa." *Cultural Anthropology* 28, no. 4: 670–93.

Von Schnitzler, Antina. 2014. "Performing Dignity: Human Rights, Citizenship, and the Techno-Politics of Law in South Africa." *American Ethnologist* 41, no. 2: 336–50.

Von Schnitzler, Antina. 2016. *Democracy's Infrastructure: Techno-Politics and Protest after Apartheid*. Princeton, NJ: Princeton University Press.

Zulu, Nobukhosi. 2018. "Access to Information Network." https://www.wits.ac.za/media/wits-university/faculties-and-schools/commerce-law-and-management/research-entities/cals/documents/programmes/rule-of-law/resources/190926_ATI%20Network%20Shadow%20Report%202018.pdf.

CHAPTER 6

Human and Machine Concept Possession

Helen Robertson

1   Introduction

Creditworthy, non-creditworthy. Malignant, benign. Likely to reoffend, not likely to reoffend. Increasingly such classification is done computationally—"by machine". Available to the machine is a large dataset with records labelled as instances or non-instances of the class of interest. Creditworthy, non-creditworthy, creditworthy, and so on. After some specified degree of accuracy is attained in classifying creditworthiness or malignancy or recidivism on the basis of the further features of the records, the machine—or more precisely, the machine learning model—is said to have "learnt the concept." Having learnt the concept, the model can then be used to classify further, concrete, instances. Miriam Joseph is sentenced to prison. Luyanda Moyo gets a loan. Mo Patel is diagnosed with a malignant tumour and is informed that he has six months to live.

The above is stated simplistically. The ways in which such classificatory models play a role in decision-making nowadays are complex. There is increasing sensitivity to the dataset and its appropriateness, both epistemically and ethically. It is, for example, commonly recognised that such classification can perpetuate injustice, if the dataset on which the classification is based itself describes an unjust state of affairs (see, for example, discussions by Hajian et al. (2016) and Hagendorff (2021)). Similarly, there is debate over the replacement of interpersonal exchange with the impersonal decision-making based on such classificatory models (for discussion of this within the Kenyan context see the chapter by Emma Park and Kevin Donovan in this volume). Furthermore, decision-making is informed by such classification in different ways and to different degrees. Within financial contexts, for example, it is common for a model to classify in a binary way (an applicant is either creditworthy or not in relation to a particular loan application) and for a decision to be based solely on this classification. In judicial contexts, by contrast, classifications tend to be more fine-grained (a convicted person is classified according to a ranking or score of likely recidivism), and to feature as one among a number of considerations that a judge will take into account in reaching a decision. Similarly, in medical contexts, practitioners will consider the classification generated by such a model as one of multiple factors that inform a final diagnosis. Nevertheless, the role of such classification, classification on the basis of a machine having "learnt the concept," is increasingly prevalent in cases in which a human—typically an expert—would previously have done this classificatory work.

In the cases in which human beings do this classificatory work, there is, both historically and presently, great variation in the circumstances, languages, practices, and behaviours that accompany such classification. The oncologist in the United States, for example, is constrained by complex health insurance regulations, with a diagnosis being captured on a large database system integrated with a patient's insurance benefits and other membership programmes. The oncologist at a public hospital in Johannesburg will perhaps examine many more patients per day than their American co-practitioner, with a diagnosis discussed with the patient in more than one of the country's many official languages.

Nonetheless, we take these human experts to be capable of such classification across these differences in circumstance, practice, and behaviour. We take the oncologist in the United States and the oncologist in Johannesburg to be competent at classifying instances of malignant and benign tumours. Part of what characterises these experts across this variation is that they can be said to possess the relevant concept(s)—and indeed to possess the same concept(s). Though the oncologist in the United States says, "malignant tumour" and the oncologist in Johannesburg says, "*isimila esibi*," their terms are translatable. The two oncologists possess the same concept of a malignant tumour and the instances that they classify are instances of this concept. The relevant members of staff at financial institutions, whether in the Northern or Southern Hemisphere, possess the concept of creditworthiness. Judges deciding on the parole of an offender in Borneo and in Japan possess the concept of recidivism.[1]

Precisely what such concept possession consists in and how it relates to classification are substantive questions. Supposing, for the moment, an intuitive understanding of "concept," possessing a concept is a necessary, though not sufficient, condition for the expert's correct classification of instances of the concept. That is, such classification is not possible without possession of the concept, but might well require more besides. Intuitively, an oncologist who does not possess the concept of a malignant tumour cannot diagnose one when she or he comes across the relevant lung X-ray, even if they can visually distinguish the relevant X-rays from one another. Similarly, a judge who does not possess the concept of recidivism cannot take likely recidivism into account as a consideration in deciding whether to grant parole.

---

1   In the case of the concepts of creditworthiness and recidivism, claims of sameness or identity of the concept are, admittedly, complicated by the partial relativity of the concept to the relevant financial and legal frameworks.

In the case of classification by machine, as noted, we equally find mention of concepts: Once a machine learning model attains some specified degree of accuracy in classification, it is said to have "learnt the concept." If, for example, a model attains the relevant degree of accuracy in classifying creditworthiness on the basis of a large historical data set of clients' transactions, it has "learnt the concept" of creditworthiness. Importantly, however, within these contexts this phrase is used, and is to be understood, in a technical sense. A model has "learnt some concept C" when it classifies Cs and non-Cs sufficiently well according to some number of specified technical measures. A machine's having learnt the concept C in this technical sense, then, is not equivalent to a machine having come to possess a concept in the intuitive sense above: The machine that classifies X-ray images of malignant tumours sufficiently well according to the relevant technical measures does not *thereby* possess the concept of a malignant tumour in the sense that the medical practitioners in the United States and Johannesburg do.[2]

In what follows, the interest will be with this point of comparison. Classification with a high degree of accuracy is possible both in the case in which the human expert does the classificatory work and in which the classification is generated by a machine learning model. In the former, we take the expert to possess the relevant concept(s). Indeed, despite the variation in circumstance, language, practice, and behaviour that accompanies such classification, and indeed despite the concerns of Willard Van Orman Quine, we take experts to possess the same concept and for this variation to be translatable by virtue of this concept. In the case in which the classification is generated by a machine learning model, we similarly find variation, and indeed greater variation, in the circumstances, "language," and behaviours accompanying this classification. The expert in Johannesburg and the one in the United States behave differently to one another, but the behaviour of the machine is yet more different to both experts, though all three—South African expert, American expert, and machine—classify with a high degree of accuracy. Is there any sense in which the machine can equally be said to possess the relevant concept(s)? That is, what is the extent, if any, to which a machine can, having attained the relevant degree of accuracy in classification, be said to possess the relevant concept(s)? And, correlatively, is there anything additional, and of significance, in the case of the expert possessing the relevant concept(s) that is not to be found in the case of machine generated classification?

---

2   Similarly, a model that attains a certain degree of accuracy in classifying trees using satellite images does not thereby possess the concept of a tree. (For discussion of such models within responses to climate change, see Véra Ehrenstein's chapter in this volume.)

In order to answer these questions, an account of what it is for a human to possess a concept is required. In what follows, we will proceed by appeal to a number of analytic-philosophical accounts of human concept possession. Some of these accounts are considered more plausible than others. For example, accounts that characterise human concept possession in epistemic terms are considered more plausible than those that characterise it in behavioural terms. Nonetheless, in so proceeding, it will become apparent that, even according to the most uncommitting account of human concept possession, machines that attain the relevant degree of accuracy in classification seemingly cannot be said to possess the relevant concept. Of course, such classification has its roots in the research and learning that take place within the rooms and corridors of universities, where students and researchers are not blind to such questions and comparisons.

2   Cartesianist Concept Possession

At a university in Johannesburg, at a time when the globe is slowly recovering from the peaks of a world-wide pandemic, a faint scratching sound is heard along the corridors of a building on the western side of campus. The scratching grows louder and then stops. On the same floor, students in computer science laboratories discuss an assignment. The assignment mentions a binary classification algorithm. One student works on their training data. Rabbit, non-rabbit, rabbit, non-rabbit... A faint scratching sound is heard. Another student awaits results. The algorithm has run for two days. The scratching grows louder. A rabbit scurries by. "Rabbit!," a student outside the lab shouts and points. The machine says, "*gavagai.*" "Ha-aaa. My results are good enough. I am going. It has learnt *rabbit.*"

A central branch of machine learning is the learning of supervised classification tasks. In such tasks, the machine learning program or algorithm is provided with examples and non-examples, labelled by humans, of the concept to be learnt. If the program is able to classify to some specified degree of accuracy, the machine is said to have learnt the concept. In the case of human concept possession, the human who shouts "rabbit!" as a rabbit scurries by is understood to possess the concept *rabbit* and their utterance to have a referential relation to the scurrying rabbit.

> Nahhh, man, the machine hasn't learnt *rabbit.* My little sister has learnt *rabbit,* but that's because she can think, man. She has some idea in her head and the idea is about rabbits and when she sees a rabbit, the idea comes into her head. And she can think about rabbits, and dogs,

and cats, and mice, and when she's thinking about them, she's thinking about them as different things. But my baby brother, he's really small. He hasn't learnt *rabbit*, because he can't think of rabbits and dogs as different things. But he will. Because he *can* think. That machine—that machine has not learnt *rabbit*. It won't ever learn *rabbit*. That machine can't think.

The taller student of the two, Siya, is an applied mathematics student who's recently joined the university from Soweto. Aside from being nicknamed "The Brain" at school, Siya was known for her scepticism. No matter what the topic,—science, religion, or even politics—Siya would not believe what anyone said without getting to the bottom of it herself—including Kabelo's claims about machines' learning concepts. Kabelo, the shorter student of the two, knows this only too well. On a scholarship from Limpopo, he is in his first year of a newly-established Masters degree programme in data science. His mother would tell him that his tongue is too quick: "*Leleme ha le na malokeletso*" (The tongue has no fastenings). Kabelo's tongue did not often have fastenings. He incites Siya's scepticism with his claim that his computer can learn: "Ha-aaa. No, no, no. You're right, man. The machine can't think, as in *think*, but it's learnt the concept. If it gets a rabbit, it mostly spits out rabbit. If it doesn't, it mostly doesn't."

According to a recent account of human concept possession, a subject *S* can be said to possess a concept *C* if he or she is "able to think about Cs as such" (Fodor 2004, 31). More specifically, thinking about Cs as such involves having a mental representation, *C*, that has a nomological or lawlike relation to Cs. Suppose, for example, that as Siya arrives to Kabelo's home to visit him over the July holiday, there is a dog running in the street. Siya thinks of an incident in her childhood and how she now dislikes all dogs everywhere. The representation in Siya's mind at this point—the representation *dog*—has a nomological or lawlike relation to dogs in general or as such. The representation, we might say, "covers" or "extends" to the dog in the street in front of Siya. It also covers the dog of the incident in her childhood, and the dog running down 7th Street in Gqeberha, and the dog that will be born exactly 102 days from now, and so on. This relation between the concept *dog* in Siya's mind and each of these dogs (and indeed all dogs) is lawlike.

This account, Cartesianism, when conjoined with most popular accounts of the mind, would seem to imply that a machine or computer like Kabelo's that has, in machine learning parlance, "learnt the concept *rabbit*" has not, in literal parlance, learnt or come to possess any concept. Most popular accounts of the mind (or the mental) do not straightforwardly deny the possibility of a machine's having mental representations of one or another sort. That is, it is compatible with these accounts that a machine could possess a mind and could think: according to functionalist accounts of the mind, for example, a machine with sufficiently rich and complex input-output functions might be ascribed mental representations of various kinds, including that of thought (see for example David Braddon-Mitchell and Frank Jackson's discussion (2007, 43–45)). Similarly, a non-reductive materialist account of the mind would allow mental states or processes to be realised in markedly different physical systems (see for example discussion by Jaegwon Kim (1992)). Nevertheless, the same accounts would deny that the contemporary machine or computer that attains a sufficiently high degree of accuracy in classifying Cs and non-Cs has come to possess any concept. The input-output functions of such a machine are neither sufficiently rich or complex nor of the relevant sort to ascribe mental representations to it. According to these accounts, Kabelo's machine cannot be said to possess the concept *rabbit*—cannot be ascribed the lawlike mental representation *rabbit*—because it cannot be ascribed any mental representation at all. Kabelo's machine does not possess a mind and cannot think.

3 Pragmatist Concept Possession

Cartesianism is not, however, the only available account of human concept possession. According to a more popular rival account, Pragmatism, a subject *S* can be said to possess a concept *C* if he or she is "able to distinguish Cs from non-Cs" (Fodor 2004, 31; see also discussion by Bradley Rives (2009a, 2009b) and Victor Verdejo (2013)). Siya possesses the concept *dog* if, when seeing the dog in Kabelo's street run after a feral cat, she can distinguish the dog from the cat (and from Kabelo the human, and so on).

Prima facie, Pragmatism would seem to imply that a machine like Kabelo's has learnt or come to possess the concept. In the running of the algorithm, the machine has come to be able to classify to some specified degree of accuracy further examples as examples of Cs or non-Cs. Successful approximation of the function just is for the machine to be able to do this classificatory work and so is *ex hypothesi* for it to be able to "distinguish Cs from non-Cs".

It is worth noting here that the machine having only attained a specified degree of accuracy need not rule out possession of the concept *C*: in the case of the person who possesses the concept *dog*, it is consistent with the Pragmatist's claim that the person fails, in some subset of cases, to distinguish dogs from non-dogs.[3] The condition for possessing the concept *C*, according to

---

3 The reason for this, as we will see below, is that the ability to distinguish Cs from non-Cs is a dispositional one. It is widely accepted that something's having the disposition to X does not

the account, is that the person is able to distinguish Cs from non-Cs in most cases. In the case of the machine that possesses the concept, classification to some specified degree of accuracy ensures that it is able to distinguish rabbits from non-rabbits in the dataset in most cases. Siya disagrees. "Ahh. I don't know if it does, Kabelo. I mean, telling rabbits from non-rabbits … There's more to this than spitting out rabbit or non-rabbit. When I tell you 'There's a rabbit over there,' there are other things that you also know because of this. You'd know that there's a mammal over there and that, if you cooked it, it'd be reee-ally tasty …"

"Hah. Ah, man. I ain't never eaten rabbit."

More specifically, according to Pragmatism, to distinguish Cs from non-Cs is not simply to be able to distinguish or classify some further examples as examples of Cs or non-Cs, but is to have a richer set of epistemic capacities: a subject S possesses a concept C if S is disposed to draw (or otherwise to acknowledge) some of the inferences that contain that concept (Fodor 2004, 33). Suppose, for example, that while talking that afternoon Kabelo tries to persuade Siya that, since Siya likes any animal that is a mammal, Siya should also like dogs. Suppose that Siya has no disposition to draw this conclusion, but that she is disposed to draw the analogous conclusions about, for example, cats. In this case, we might deny that Siya possesses the concept *dog* (though would affirm that she possesses the concept *cat*).[4]

More specifically then, Pragmatism would imply that a machine or computer like Kabelo's that has, in machine learning parlance, learnt the concept *rabbit* has not, in literal parlance, learnt or come to possess any concept. The machine can classify the further examples accurately, but it is in no way disposed to these further inferences. It is not disposed to any further inferences.

It is worth here emphasising the distinction between Cartesianism and Pragmatism. Both are accounts of concept possession according to which concepts play a role in, and thus require, thought or thinking. According to Cartesianism, to possess a concept C is to be able to think about Cs as such. According to Pragmatism, to possess a concept C is to be disposed to certain inferences, which themselves seem to require thought. The two are distinct, however, in the way in which thinking or thought is appealed to and presupposed. Pragmatism is an epistemic account and whether or not some subject

---

entail that it will not fail to X in some subset of cases. See for example Michael Fara's (2005) discussion.

4   It is to be noted that these epistemic capacities are only some of those appealed to by the Pragmatist, with different versions of Pragmatism appealing to different—and sometimes mutually-exclusive—capacities.

possesses a concept will be subject to the normative constraints on knowledge. Cartesianism, by contrast, requires only a lawlike relation between the representation C and that which the representation is about, Cs, though it is prior to and does not presuppose the subject's having knowledge of Cs.

"Woah, woah, woah, man. I thought we admitted that computers can't think. And so they can't do this reasoning thing that you're talking about. But telling rabbits from non-rabbits, learning a concept, isn't thinking or reasoning, it's being able to behave in an input-output way. Like I said, if it gets a rabbit, it mostly spits out '*gavagai*'. If it doesn't, it mostly doesn't." In defence of his machine learning model, Kabelo's tongue still has no fastenings.

4   Quinean and Type-Relative Quinean Concept Possession

A further account of concept possession that might be proposed, one far more minimal than both Cartesianism and Pragmatism, is, what we might term, a "Quinean" account.[5] According to this account, concept possession does not in any way consist in or require that the possessor of the concept possesses a mind or can think or has mental representations or draws inferences. According to our Quinean account, a subject S possesses the concept C if he or she is disposed to exhibit some set of verbal behaviours B in response to Cs. For example, if Siya is disposed to say the word "dog" when asked "what is that?" by someone who points to a dog, and to answer "no" when asked "is this a dog?" by someone who points to anything that is not a dog, and so on, then Siya can be said to possess the concept *dog*.

It is worth noting the sense in which the account is a naturalist one. The account appeals only to, what we might term, "natural" properties. Such properties are those that might be described in the various physical sciences and whether or not such a property is instantiated can be determined by observational means. For example, biological properties such as being a cell of a certain sort or physical properties such as being larger than one cubed centimetre are natural properties, while the properties of being morally good and being a mental representation are not.

The properties mentioned in the Quinean condition for concept possession similarly are those described in the physical sciences and are those whose

---

5   More properly, the Quinean account is an account of the (un)translatability of natural language and thus an account of linguistic terms, not concepts. Here, the term "Quinean" will be used loosely to refer to the naturalist account described—an account that, although not to the letter, is perhaps Quinean in spirit. See for example Quine (1960, Chapter 2).

instantiation can be determined by observational means. Concept possession is, according to the account, describable in the terms of anatomy and linguistics, and whether or not some subject possesses some concept is determinable by observing whether their anatomy is engaged in the relevant ways. I can be said to possess the concept *dog* if I am disposed to form my lips into the sounds that compose the word "dog" when asked "what is that?" by someone who points to a dog, and so on.

Under this much weaker account then, a machine like Kabelo's could perhaps be said to have learnt or come to possess the concept. As emphasised, the Quinean account is not committed to thought as a requirement for concept possession. Indeed, the account eschews all talk of the mental as internal to the subject. All that is required, according to the account, is that the machine is disposed to exhibit some set of behaviours in response to Cs. The machine does indeed, as Kabelo notes, exhibit some set of behaviours in response to Cs: if it gets a rabbit, it mostly spits out "*gavagai.*" If it doesn't, it mostly doesn't.

"Kabelo, dude. What's up with '*gavagai*'? Why does it spit out '*gavagai*'? This is not a language."

"Yo, man. A friend studying philosophy has some trippy ideas about language not being translatable and about someone talking to some native tribe and not being able to tell whether they were both talking about a rabbit. The man from the tribe said '*gavagai*' when he shoulda said 'rabbit'. That's what I changed the label to." The rabbit rushes by. The taller student picks up the rabbit and shows it to the computer.

"Na-ah, Kabelo, man. Look …! It doesn't have eyes. The machine can't distinguish rabbits." Siya shows the rabbit to the computer. "It wouldn't be able to point one out even if the rabbit was sitting on top of it …!" She puts the rabbit down on top of the computer on which Kabelo has run his code. The rabbit stares ahead.

Examining the Quinean account more closely, there might indeed be reason to suppose that the machine does not exhibit a relevant set of behaviours in response to rabbits. I can be said to possess the concept *dog* if I am disposed to form my lips into the sounds that compose the word "dog" when asked "what is that?' by someone who points to a dog, to answer "no" when asked "is this a dog?' by someone who points to anything that is not a dog, and so on. A contemporary computer or machine has no lips and so it cannot form its lips into the sounds that compose the word "dog" when asked "what is that?" by someone who points to a dog. And so the machine cannot be disposed to form its lips into the sounds. Given this then, the account seems equally to imply that a machine that has been trained on certain data and that has come to "learn the concept *C*" has not learnt or come to possess the concept *C*.

Ha-aaa. No, man. You can't put that on the computer! You can't say that the computer doesn't recognise the rabbit because it doesn't have eyes and ears and it can't walk and talk and all that. *It's a computer.* Of course it don't have eyes and ears. It distinguishes *in its own way*, man. It doesn't have eyes, but it's got a processor that can run my algorithm and it 'looks' at the data with these eyes. And it 'tells' us that it's a rabbit by an output that a computer can have man. Its screen is its mouth. You can't be putting that on the computer man. Trying to force it to be human.

Kabelo grows louder. Some students passing by put their heads in the door of the lab to investigate the commotion. Kabelo's rejoinder here then is to appeal to a type-relative (or species-relative) account of behaviour: although most humans are disposed to the verbal behaviours described above, those who are deaf, for example, would not be. A person who is deaf would not be disposed to the behaviours of forming their lips in the relevant ways and so on, but instead to certain gestures with their arms and hands. Yet in the case of the human with differing capacities, we do not thereby suppose that the human does not possess the concept, but instead take a different set of behaviours to be relevant to whether they do.

So too, continues the response, the computer or machine has different capacities to a typical English-speaking human for responding to rabbits and non-rabbits, and the behaviours to which a machine must be disposed in order to count as possessing the concept will be relative to these capacities. Whether or not a machine possesses some concept C will depend on whether it is disposed to exhibit some set of behaviours B in response to Cs, where the behaviours B include certain screen outputs given certain data inputs. According to this type-relative version of the Quinean account then, a machine that is in this way disposed can be said to possess the concept.

5   The Problem of Generalisability

"Yeah... Sure. The computer doesn't have eyes and it displays it in another way. But it's still limited, man. It only has the right dispositions to act in a small range of cases. Give it the dataset. Sure, once it's trained, it'll tell rabbit from non-rabbit in the test data. But it can't tell me that *this* is also a rabbit." Siya points to the rabbit, which is now sitting underneath the lab desks.

"Nah, man. That's the problem of generalisability. It's covered. We got it. There's some theory that shows it's not just for one dataset. Look, what the machine is doing is approximating some function. It's approximating some real function. The machine can tell all rabbits from non-rabbits, man. If you included the values of this rabbit in the dataset, it'd tell you." Kabelo points to the rabbit under the desks. The rabbit's nose quivers.

As seen above, according to the type-relative Quinean account, a machine that is able to classify further examples as examples of Cs or non-Cs to some specified degree of accuracy is disposed to some relevant type-relative behaviours and so can be said to possess the concept. (The machine has a certain screen output when given certain data inputs.) Having examined the various ways in which the capacities of the machine might prevent it from possessing a concept, and having arrived at the possibility of the machine having some relevant capacities, once these are understood as type-relative, a further set of concerns arises. Admitting some relevant type-relative behaviours, the concerns are over whether the machine is disposed to exhibit the behaviours across a sufficiently wide range of cases.

In the example above of Kabelo's running a binary classification algorithm, the algorithm has allowed the machine to identify (or rather, approximate) some function or rule for classifying the labelled examples into Cs and non-Cs, with the machine thereby disposed (with some probability, that is, to the specified degree of accuracy) to exhibit the relevant type-relative behaviour in response to some further examples (the test data). The concern here, however, is the distinction between the machine being disposed to some type-relative behaviour in response to further examples of rabbits and its being disposed to exhibit some type-relative behaviour in response to *any further examples* of rabbits.

In the case of Siya's possessing the concept *dog*, Siya will be disposed to form her lips into the relevant sounds when faced with the neighbour's German Shepherd, when faced with her mother's Fox Terrier, when faced with a Bull Mastiff in some suburban area while on holiday, and so on *ad infinitum*. That is, possession of concept *C* requires the disposition to exhibit some type-relative behaviour, not only in response to some further subset of Cs, but all actual and possible Cs.[6] As it stands, there is no reason to suppose that a machine like Kabelo's is so disposed.

In order to answer the concern, we must, as Kabelo rightly indicates, turn to the theory of computational and machine learning, that is "learning theory." In particular, we must turn to the Probably Approximately Correct model of

---

6   Again, the requirement is a dispositional one and is compatible with the human failing to do
     so in some subset of cases.

learning, a model that constitutes a general account of binary classification machine learning algorithms.[7]

6   Probably Approximately Correct Learning

The Probably Approximately Correct model (hereafter, the "PAC model") of learning distinguishes a subclass of tasks that are learnable within specified computational constraints. That is, it distinguishes a subclass of tasks for which a machine learning program or algorithm is possible assuming some upper threshold on the number of examples and time and space complexity allowed for the computation. Taking Kabelo's case as our example, the subclass is characterised as follows.

Available to the machine learning program is (1) the set, S, whose members are examples of mammals, where each example is a set of values for the features of colour (represented as a colour code), height (in metres), and region (represented as an area code), (2) a label for every member in S as either "non-rabbit" or "rabbit," and (3) a set of hypothesis functions H. This set of hypothesis functions is a set of (mathematically-defined) possible rules that would allow the program to "go from" the values in the set S to the labels in (2). The task of the program is to select, on this basis, the function or rule that best maps the examples to their labels. In other words, from this set of possible rules, the program is to select the rule that allows it to classify correctly as many of the examples, as rabbits and non-rabbits respectively, as possible.

Presupposed in the background, however, is that the function or rule to be selected by the program is not simply the function from the given finite set S of examples to their labels. Instead, the program is to select the rule that best approximates the more general function or rule from all possible such examples to their respective labels. In other words, Kabelo's machine is to select, from the possible rules available to it in H, not simply the rule from the examples of mammals in the data set to the examples of rabbits in the data set. Instead, it is to approximate the rule from all possible examples of mammals to all possible examples of rabbits. If it is able to do so, then the machine will be able to classify not only the examples of rabbits found in the data set, but, as

---

7   The PAC model of learnability was first presented in Leslie Valiant (1984). Recent well-known expositions of the model that fall in line with the exposition below include discussions by Ethem Alpaydin (2004), Tom Mitchell (1997), and Mehyrar Mohri (2012). Later versions of the model extend the account to non-binary and unsupervised learning tasks. Discussion of the primary versions of the model are adequate for our purposes.

Kabelo rightly responds, it will also be able to classify the rabbit that now sits underneath the lab desks.

Presupposed in the background of such learning then are also the following. (4) The larger set, X, the members of which are all possible examples, each itself a set of feature values, and of which S is a subset. In Kabelo's case, this set is the set of all possible examples of mammals. The examples that appear in Kabelo's data set form a subset of these examples and are the set S as introduced in (1) above. (5) A label for every member in this larger set X as either "non-rabbit" or "rabbit." (6) The function or rule from the members of the larger set X to their respective labels. In machine learning parlance, this function or rule is known as the "target concept c." Importantly, this function or rule is not one of the possible rules that is found in the set H, from which the machine can select. It is the actual or true function or rule that allows us to go from all possible examples of mammals to all of the examples that are rabbits. Knowing this function would allow one (or Kabelo's machine) to classify every possible example of a mammal correctly as either a rabbit or non-rabbit.

This rule is, of course, unknown to Kabelo's machine and the task of his machine is to select, from the set of hypothesis functions H, the function that best approximates this true function. If it is able to do so sufficiently well, then the machine will have achieved the relevant degree of accuracy in classifying rabbits and non-rabbits and will, in machine learning parlance, have learnt the target concept c.

Supposing the Probably Approximately Correct model (PAC) to be the correct account of binary classification algorithms, it would seem to imply that Siya's concern is unfounded. Recall, Siya's concern is over whether the machine is disposed to exhibit the behaviours across a sufficiently wide range of cases: can the machine correctly classify, not only the examples in the data set that Kabelo has given it, but all further such examples?

As we have just seen, according to the PAC model, the target function for the machine learning program is not simply the function or rule from the set of examples of mammals, as made available by Kabelo in running the algorithm, to the subset of those examples with the label "rabbit". It is rather the more general function from all possible examples of mammals to the subset of those examples with the label "rabbit" (that is, the subset of rabbits). Once this function has been approximated, Kabelo's machine *is* disposed (with some probability, that is, to the specified degree of accuracy) to exhibit the relevant type-relative behaviour in response to *any further examples*.

As Kabelo rightly notes, the mentioned concern is a concern over generalisability: does the accuracy of the machine in classifying extend sufficiently beyond just the dataset on which it has been trained? The PAC model of learnability constitutes a mathematical account of such generalisability and allows us to answer the question in the affirmative.

7   The Problem of Generalisability, Again

"Yeah, yeah. I've heard of this. But it's not *all rabbits at all times everywhere*, Kabelo. It can't know about the rabbits that will exist in the future and that will look weird and different. But we will. We'll be able to tell that they are rabbits."

Siya's concern runs a little deeper. Recalling the type-relative Quinean account in the case of human concept possession, supposing that a human were to possess the concept *rabbit*, the human would exhibit the relevant type-relative behaviours in response to examples of rabbits—even in cases in which the rabbits were radically different. Supposing that Kabelo possesses the concept *rabbit*, Kabelo would exhibit the relevant type-relative behaviour in the case of the rabbit underneath the desk, of the rabbit on the farm in Limpopo, and equally in the case of the rabbit that has been selectively bred to have fur much longer than most rabbits. The further concern is whether the machine would be able correctly to classify even in these cases. Kabelo might have shown that the machine can classify the rabbit underneath the lab desks correctly, but can it correctly classify the rabbit that exists one hundred years from now and which has been selectively bred to stand over one metre in height?

We might suppose that Kabelo has already answered this. After all, according to the PAC model of learnability, the target function c, the function that machine learning program is trying to approximate, simply is the function from the set X of all possible examples to the subset with the relevant label: it is the function for classifying all possible examples of mammals as rabbits or non-rabbits, including the one-metre rabbit that exists one hundred years from now.

Appeal to the PAC model of learnability is not quite so straightforward, however. Two formulations of "all possible examples" can be found in descriptions of the model. That is, descriptions of the model seem to be ambiguous as to what is meant by "all possible examples." The first formulation is found in informal descriptions of the model and is broader in scope, while the second appears in the model's formal description and is narrower in scope. Let us begin with the former.

In descriptions of PAC learnability, the model is frequently illustrated informally with an example of a simple learnable task. Such an illustration is found, for example, in Tom Mitchell's discussion: "[L]et X refer to the *set of all possible instances* over which target functions may be defined. For example, X might represent *the set of all people*, each described by the attributes age (e.g., young or old) and height (short or tall). Each target concept c [...] corresponds to some subset of X, or equivalently to some boolean-valued function $c : X + \{0, 1\}$. For example, one target concept c might be the concept '*people who are skiers*' (Mitchell 1997, 203).

Illustrations like Mitchell's suggest that the set X here is to be understood as all possible instances of some higher-order kind and the subset of examples with the relevant label as the set of all possible instances of the relevant lower-order kind. The superset X might be, as in Kabelo's case, the set of all possible instances of the higher-order kind *mammal* and the subset all possible instances of the lower-order kind *rabbit*. Or the superset X might be, as illustrated by Mitchell, the set of all possible instances of the kind *person*, with X' the subset of all possible instances of the lower-order kind *skier*. Similarly, X might be the set of all possible instances of the kind *animal* and X' the set of all possible instances of the kind *bird* (Mitchell 1997, 20–21). Thus, according to the first formulation, "all possible examples" are all possible instances of some (higher-order) kind and the subset of examples with the relevant label are all possible instances of the relevant lower-order kind.

According to this first formulation then, the scope of "all possible instances" is determined by the higher-order kind. Because this is so, a machine that approximates the function from all possible examples (that is, all possible instances of the higher-order kind) to all possible examples with the relevant label (that is, all possible instances of the lower-order kind) would indeed be approximating a rule that allows it to classify all possible examples—in Siya's sense. Given that it includes all possible instances of the higher-order kind (and thus all possible instances of the lower-order kind), the set X would include examples that are radically different. A machine approximating this function would thus be able to classify, to some specified degree of accuracy, all possible examples. It thus would be disposed to exhibit the type-relative behaviour in response to any further examples—even to the rabbit in one hundred years' time that stands at one-metre tall.

According to the second formulation found in more formal descriptions of the model, we find that "all possible examples" is to be understood as all possible feature-value combinations, given the features and possible values of the learning in question. To clarify, it will be helpful to return to Kabelo's machine and, in particular, to the data set that has been made available to it. As noted in the preceding section, available to the machine is the set S, whose members are examples of mammals, where each example is a set of values for the features of colour (represented as a colour code), height (in metres), and region (represented as an area code), with the entire set represented as a single value or vector. In simplified form, the set S might thus consist in the following.

TABLE 6.1   Simplified version of the set S in Kabelo's binary classification task (Author's own)

Example	Colour (where B = brown, Bl = black, W = white)	Height (m)	Region (where SA = South America, A = Africa, E = Eurasia, Au = Australia)	Rabbit
1	B	0.25	SA	Yes
2	W	0.45	E	No
3	Bl	0.305	Au	Yes
4	Bl	0.29	A	Yes

Important to note here is that each of the members of the set S—each of the examples above—is itself a set of values for each of the features of colour, height, and region. Example 1 is the set {B, 25, SA, Yes}. Example 2 is the set {W, 45, E, No}. And so on. It is these sets of values that are available to Kabelo's machine.[8]

As noted, and presupposed in the background of the PAC model, the set S above (of some possible examples of mammals) is a subset of the set X (of all possible examples of mammals). In order to answer Siya's concern with the generalisability of the machine's learning capacity, we noted that the target function—the rule to be approximated by Kabelo's machine—was the function from this superset X (of all possible mammals) to the subset of all possible mammals that are rabbits. If the set S is a set like that characterised above (a set of feature values), what then would it be for X to be a superset of all possible such examples? When stated in this way, it is unnatural to suppose that X would be a set of all possible instances of some higher-order kind as understood above. Instead, given that the set S is a set of feature values, it is natural to suppose that the superset of all possible such examples is the set of all possible combinations of values for the given features. In the case of Kabelo's machine, the set X is the set of all possible distinct combinations of the values

---

8   For discussion of the contrasting forms that some of these values would take within the Yorùbá numbering system, see Helen Verran's chapter within this volume.

for the features of colour, height, and region. Only some of these combinations appear in the subset S, but the superset X is the set of all possible such combinations. Thus, according to the second formulation, all possible examples are not all possible instances of some higher-order kind, but are all examples with some distinct possible feature-value combination.[9]

Under the first formulation then, the function to be approximated by the machine is one from all possible instances of a higher-order kind to a subset of all instances of some lower-order kind. Under the second formulation, the function to be approximated is from the set of all possible combinations of feature values (the set of all possible colour-height-region combinations) to the subset of all such combinations to which the label can be assigned (the subset of all colour-height-region combinations applicable to rabbits).

It is important to note that the two formulations above are not equivalent in all cases. In cases in which the values for any feature are underestimated in relation to all possible instances of the higher-order kind, the first formulation will be broader in scope than the second. If, for example, Kabelo underestimates the range of heights of all possible rabbits that have existed and do and will exist, the set of all possible instances of the lower-order kind rabbit will be broader in scope than Kabelo's possible combinations of feature values will cover.

If the two formulations are not, as suggested, equivalent, then the PAC model must presuppose one or the other. It is reasonable to suppose that this is the second formulation, for two reasons. First, the second formulation is the formulation found in the model's formal descriptions. Secondly, while many illustrations cohere with the first formulation, equally many do not (see for example Mitchell's discussion (1997, 22)). What is the implication of this then for Kabelo's claim that his machine is able to classify any further examples of rabbits?

As noted above, the first formulation seems to do the work needed for the Kabelo's response. If the superset X is the set of all possible instances of some higher-order kind (the set of all possible mammals), this would ensure that the function approximated is to the set of all possible instances of the lower-order kind (the set of all possible rabbits). This would then ensure that the machine is disposed to exhibit some type-relative behaviour in response to any further examples. What of the second formulation? Does the second formulation similarly ensure that the machine is so disposed?

Recalling the inequivalence mentioned above, the second formulation would seem to imply the following. According to the formulation, the machine approximates the function or rule from the set of all possible combinations

9   In the case of complex and high-dimensional data, these combinations can be extremely large in number. This does not itself affect the point made above.

of feature values (the set of all possible colour-height-region combinations) to the subset of all combinations to which the label can be assigned (the subset of all colour-height-region combinations applicable to rabbits). Now whether this would be to approximate a rule that would allow it to classify all possible examples—in Siya's sense—will depend on the range of features and values that are found in those combinations. Whether Kabelo's machine would be able to classify all possible examples of rabbits would depend on the features found in the data set (the features of colour, height, and region) and the possible values for the features that Kabelo programs the machine to recognise. If the features in the data set are indeed features of the relevant examples, and if the possible values for these features exhaust their values for all possible such examples, the machine will be able to classify—to some specified degree of accuracy—all possible examples. That is, if the features found in Kabelo's data set (of colour, height, and region) do indeed characterise rabbits, and if the possible values that Kabelo programs the machine to recognise for these features (the values of brown, black, white, of twenty-five to forty-five centimetres, and of the continental regions) exhaust the range of values for all possible examples of rabbits, then Kabelo's machine will indeed be able to classify all possible examples of rabbits. And, in such cases, Kabelo's machine will be disposed to exhibit some type-relevant response to any further examples.

If, however, either the features found in the data set are not features of the relevant examples (if Kabelo has, for example, included values for the feature of feather type) or the possible values for these features do not exhaust their values for all possible examples (if Kabelo has, for example, included only brown and black as possible values for colour), then Kabelo's machine will not be able to classify all possible examples. The function approximated by the machine will not be a function to all possible examples, but instead a function to all examples, given our current knowledge or the chosen limits of the values of the features. Kabelo's machine will classify only those examples that fall within the range of possible feature-value combinations that Kabelo has programmed it to recognise. And, in such cases, the machine would not be disposed to exhibit some type-relative behaviour in response to any further examples.

If correct, this would seem to imply that Siya's concern is perhaps founded. There is a contrast to be drawn between the dispositions to type-relative behaviour in the case of the possession of a concept by a human and in the case of the learning of a concept by machine. In the case of the latter, whether or not the machine can be said to possess the concept hinges on our exhausting the range of features and possible values of a feature, and thus on our knowledge at a given time of those values. It depends on Kabelo's making sure that the values for the feature of height range from twenty-five centimetres to the one-metre height of the rabbit that lives one hundred years from now. In the human case, however, concept possession is hardier in the face of epistemic limitations. Siya succeeds in possessing the concept *dog* or *rabbit* or *mammal*, and is thus disposed to exhibit certain behaviours in response to any further instance of a dog or rabbit, despite her lack of knowledge of the values of the features of all dogs or all rabbits and despite her not encountering the one-metre rabbit that will exist one hundred years from now.

"Well of course it's not all rabbits at all times everywhere, Siya. What do you think? That the machine is human?"

8   Conclusion

It is perhaps unsurprising that, according to stronger or more committed accounts of what it is for a human to possess a concept, the machine that has attained the relevant degree of accuracy in classification cannot be said to possess the relevant concept. The medical practitioner in the United States or in Johannesburg has, according to Cartesianism, a mental representation with a lawlike relation to all possible cases of malignant tumours, whether or not these are cases that the practitioner has encountered or will encounter. There is nothing in the machine's computations that corresponds to such a mental representation. The judge in Borneo or in Japan is, according to Pragmatism, able to reason about likely recidivism. He or she is disposed and able to draw the basic inference from likely recidivism as a characteristic of human beings to likely recidivism as a characteristic of some, but perhaps not all, mammals. The machine that has attained the relevant degree of accuracy in classification does not have these abilities.

Yet, as we have seen above, even under the most uncommitting accounts of human concept possession, accounts that do not require anything like a mind or mental representations, the machine that has, in machine learning parlance, learnt the concept, has not, in literal parlance, come to possess the concept. According to, what we termed, a type-relative Quinean account, the practitioner in Johannesburg possesses the concept of a malignant tumour insofar as she or he is disposed to certain behaviours when encountering instances of a malignant tumour. The practitioner is, for example, disposed to use the word "malignant" and to further examine the tumour in certain ways. This is true even in cases that fall outside the range of familiar cases. The practitioner is disposed to behave in these ways to any further examples. By contrast, the machine that has learnt the concept of malignancy is not so disposed. Even granting generalisability, the machine cannot exhibit the relevant classifica-tory behaviour in response to examples that differ too considerably from the examples that the machine has been taught to recognise.

So, while we can be sure that two human experts, in spite of differences in circumstances, language, practices, and behaviours, and indeed in spite of the concerns of Willard Van Orman Quine, possess the same concept, we cannot be sure that the machine, in spite of its classificatory behaviours, possesses any concept the same as ours at all.

Bibliography

Alpaydin, Ethem. 2004. *Introduction to Machine Learning*. Cambridge, MA: MIT Press.

Braddon-Mitchell, David and Frank Jackson. 2007. *Philosophy of Mind and Cognition*. Malden: Blackwell Publishing.

Fara, Michael. 2005. "Dispositions and Habituals." *Noûs* 39, no. 1: 43–82.

Fodor, Jerry. 2004. "Having Concepts: a Brief Refutation of the Twentieth Century." *Mind and Language* 19, no. 1: 29–47.

Hagendorff, Thilo. 2021. "Linking Human And Machine Behavior: A New Approach to Evaluate Training Data Quality for Beneficial Machine Learning." *Minds & Machines* 31, 563–93.

Hajian, Sara, Francesco Bonchi, and Carlos Castillo. 2016. "Algorithmic Bias: From Dis-crimination Discovery to Fairness-aware Data Mining." *Proceedings of the 22nd ACM SIGKDD International Conference on Knowledge Discovery and Data Mining (KDD '16)*: 2125–26.

Kim, Jaegwon. 1992. "Multiple Realization and the Metaphysics of Reduction." *Philoso-phy and Phenomenological Research* 52, no. 1: 1–26.

Mitchell, Tom. 1997. *Machine Learning*. New York: McGraw-Hill.

Mohri, Mehyrar. 2012. *Foundations of Machine Learning*. Cambridge, MA: MIT Press.

Quine, Willard Van Orman. 1960. *Word and Object*. Cambridge, MA: MIT Press.

Rives, Bradley. 2009. "The Empirical Case Against Analyticity: Two Options for Con-cept Pragmatists." *Minds & Machines* 19, no. 2: 199–227.

Rives, Bradley. 2009. "Concept Cartesianism, Concept Pragmatism, and Frege Cases." *Philosophical Studies* 144, no. 2: 211–38.

Valiant, Leslie. 1984. "A Theory of the Learnable." *Communications of the ACM* 27, no. 11: 1134–42.

Verdejo, Victor. 2013. "The Rationalist Reply to Fodor's Analyticity and Circularity Chal-lenge." *Theoria* 28, no. 76: 7–25.

CHAPTER 7

The Dual Metrics of Contemporary Yorùbá Life

Helen Verran

1   Introduction

Metrics are systems of measurement that facilitate quantification in various forms of valuing and ordering in everyday life, speeding up communication and enhancing mutual trust in transactions. The term metric goes back to the word *metron,* which was meaningful in ancient Greece, where it concerned the craft of both poets and geometers. In poetry it was patterns in timing of performance, while in geometry it was patterns in spatiality that mattered. Metric as I use it here is a general word implying measures in action, an imagined mesh embedded within expressed patterns of sociomaterialising relations. Metrics are patterned connections and separations in society where social relations are mediated through involvement with the stuff of the physical world. In contemporary Yorùbá life the traditional Yorùbá metric is in widespread common use. Arithmetically, linguistically, and historically, this metric is very different than the international base ten, or decimal standard metric which permeates equally widely.

The paper opens with an ethnographic story of a lively arithmetic class involving both metrics. This setting has me homing in on metrics as they come to life in classrooms where children are schooled in the rules for using so-called natural numbers, rules for valuing in counting and measuring, and rules for manipulating number relations. We are not concerned here with statistical numbers, or economistic numbers (Verran 2015). Then, having delved into the different arithmetics and linguistic practices embedded in the workings of the decimal and Yorùbá metrics, I sum up the first part of the paper by going back to the opening ethnographic story, trying to put myself into the shoes of the teacher who designed and implemented the classroom lesson I describe. This exercise of imagination has me proposing both metrics as established linguistic-arithmetic lacings that thread as supportive dynamic mesh through the sensible sociomaterial everyday we negotiate in our ordinary lives. In this picture the numbers mediated in and expressing the patterns of metrics, dynamic meshed sets of relations, happen in ways that can be felt or experienced in a here and now. I go on to develop this working imaginary using comments from children bilingual in Yorùbá and English. The commentary of these remarks by children offer descriptions of the different "feel" the between numbers by which the metrics are worked. As materialised linguistic-arithmetic entities the numbers that express the Yorùbá metric can be felt as different in use than modern decimal numbers.

I propose this different feel the children talk of as analogous to the different feels of iconic and indexical numbers (Verran 2010). I suggest that in use Yorùbá numbers as taught in classrooms lean towards iconicity—a rationality relating wholes and parts, whereas decimal numbers seem to be naturally indexical, expressing a rationality involving the one and the many. I have previously suggested that while indexical numbers are important in scientific knowledge making, but when it comes to trade, it is iconic numbers that matter. Thus, I am proposing that the metrics differ not only in the rationalities of their making in linguistic and arithmetic practices, but also in the rationalities of the practices of their primary uses beyond the classroom. This is not to deny that with slight modifications the numbers of each of the metrics as taught in classrooms can be repurposed to work the alternative rationality.

Alternative patterns of relations are embedded in the dual metrics of contemporary Yorùbá life, and these different patternings in making are felt in use. One pattern of relations begins with precisely defined ones or units, and indeed has purloined the word metric as its basic unit of spatiality: "the metre". The pattern of relations embedded in this set of numbers comes to life most strongly using English language and English numbers. It begins in precise specification of unit. The other set of numbers, embedding a different pattern of relations comes to life in Yorùbá language. This form of quantification commences with a vague whole and achieves precise definition of units of measure in processes of material partition; here definition of unit of measure is an outcome.

Arithmetically speaking, decimal numbers are based around ten, the number at which the basic set of numbers starts to repeat with some small modification in naming to indicate the number of repeats of the set, so for example, "twenty-one" starts the third repetition of the basic set. This is the internationally accepted system. In contrast, arithmetically the Yorùbá metric works through using three base points: twenty, ten, and five. While it is possible to use a base ten system when using Yorùbá (invented in the 1960s), this is fairly rare. When speaking or writing Yorùbá, it is usual to use the triple base metric, and indeed algorithms have been devised to enable computer generated text-to-speech and machine translation of Yorùbá text that includes use of the Yorùbá number system (Akinadé and Ọdéjọbí 2014).

This chapter then offers a very short introduction to the duality of the contemporary Yorùbá metrics, drawing on insights developed in *Science and an*

African Logic (Verran 2001). It understands numbers as simultaneously wordy and relational, and as materialising in socially different ways. Assuming that readers are familiar with the parts and the workings of the base ten metric I use it to offer a description of the metric of Yorùbá language. My telling assumes the two metrics can be, and are routinely, worked as connected in everyday Yorùbá life. In beginning, I note that this assumption of *in situ* connectability that is inherent in my analysis, is controversial among scholars who study the concept of numbers, and this is true both for scholars of Yorùbá number, and for scholars of the base ten international standard metric. Here I do not offer a scholarly justification of my framing which takes the possibility of intimate, albeit partial, connection of the metrics and the numbers which circulate in them, for granted—such argumentation is beyond the scope of the chapter. In the first section, I tell an ethnographic story of meeting the Yorùbá and decimal metrics in action together and separately in an arithmetic lesson. This story anecdotally helps explain why I came to adopt this frame and to seek insight into the different sociomaterial expressions of the dual metrics. We see a fairly ordinary class of children skilfully working the two metrics in intimate tandem—translating between them. I imagine them bouncing skilfully between the alternative meshes of the two metrics that thread through their ordinary lives.

Having seen and heard how bilingual Yorùbá children easily connect the two metrics in practice in lessons, I was intrigued. I went on to talk to around seventy children bilingual in Yorùbá and English asking them to tell me how they actually used the two systems when they were doing the practices involved in measuring and counting. This very complex study which can be understood as a form of conversational ethnography, involved providing actual substances to be measured and counted during discussion lasting thirty to forty-five minutes. In all, I conversed one-to-one with nearly two hundred and fifty children, some in English and some in Yorùbá. Of the seventy children bilingual in Yorùbá and English, I spoke with thirty or so of these children in English, and the other thirty plus in Yorùbá (Verran 2001).

To sum up quickly the take-away message embedded in the answers these bilingual children gave me, they said that when they were actually using the Yorùbá number system in counting and measuring, they really needed to pay attention to the purpose of the quantification, that would tell them what unit of measure to use; it is people doing particular things who make-up the units of the material being counted or measured. The sort of unit adopted in using the Yorùbá metric is based on the social reason that has people needing to measure or count. To say that more formally, the children emphasised purpose in number use as important when using the Yorùbá system. In contrast when they used English numbers they said, you need to really look at the substance, it is the "stuff" itself which determines the unit to use. As the children told it, it is physical aspects that are emphasised in using English language numbers.

I come back to this very general difference the children discerned after we have had a look at arithmetic and linguistic mechanisms embedded in the metrics. Inquiring into the different mechanisms involved in articulating the metrics we will see that the difference the children identified is associated with another other significant difference. The pattern of relations embedded in and expressed by Yorùbá numbers foregrounds a whole-parts trajectory in use, so particular usage begins with some sort of vague whole and ends up with precise but distinct relational parts, which acting backwards together define the beginning whole precisely. Whereas the standard international decimal numbers foreground a one-to-many relation, which in use begins in a precise one and ends up with a many which projecting beyond and outside the situation of quantification, becomes a precisely defined whole ready to be reconstrued as a precise one in a further context. Thus, there is a sense of abstraction, a carrying off to another domain associated with decimal numbers and the modern metric.

2   Contemporary Yorùbá Metrics in a Classroom Setting

The school is located on one of the narrow streets radiating off from the Oòni's palace Ilé ifẹ̀ in Nigeria, the roadway usually clean and relatively clear of cars. The school grounds too, much smaller than most schools, always seemed neat, rake marks showing in the dust when we arrived in the morning. Mr. Ojeniyi is dressed as usual in a clean, crisply ironed white shirt. The older children in their final years of elementary education are quiet and responsive to their teacher sitting in orderly rows, their exercise books open before them. I always enjoy Mr. Ojeniyi's lessons, he is one of those people turned on by the aesthetic of math. Numbers and their relations clearly provoke joy and pleasure for him, which he communicates easily to the children. His lesson concerns the translation between the base ten numbers of English language metrics, and the complicated multi-base numbers of Yorùbá, which pivot around twenty, ten and five.

Mr. Ojeniyi begins his lesson in English with the statement, "You will not understand a number unless you understand the many ways it can be divided". This is certainly not the usual way that a whole number is understood in primary school mathematics. After a few sentences Mr. Ojeniyi shifts to Yorùbá as he gets warmed up in his explanation. I close my eyes to try to pick-up the Yorùbá better and to follow his reasoning. I keep losing his line of argument in my increasing agitation over the unorthodox account of what a number is that Mr. Ojeniyi is articulating; the Yorùbá words are becoming opaque. I try to calm myself and just listen. Looking around, it is obvious from the looks on the children's faces both that they understand and that they are keenly interested in what Mr. Ojeniyi is saying. I am the only one in the classroom not paying full attention to the teacher.

Relaxed and catching again the gist of Mr. Ojeniyi's reasoning: he names a Yorùbá number, too complicated for me to hear. Then he explains it as a factor of twenty, plus or minus various factors of twenty, I remember these Yorùbá number names well enough to follow his account. Then helpfully, he translates it into a base ten English language number and plots out the arithmetic process traced in making the Yorùbá number using English number names and the base ten system. Then he starts again with an English language number name and converts into a Yorùbá number, using division into sets of twenty as the first and defining process. A complicated explanation switching between Yorùbá and English language in a precise, comparative performance of two very different numbering systems.

The children had clearly all followed his explanation. They then copy down in their exercise books the series of arithmetic operations Mr. Ojeniyi has elaborated for each translation on the blackboard. After two more such demonstrations are duly spoken and inscribed on the blackboard—one a Yorùbá number translated to English base ten number, the other an English number translated to Yorùbá—Mr. Ojeniyi names a number in English, "Nineteen thousand, six hundred and sixty nine" and writes "19,669" on the blackboard. How can this be named in Yorùbá? Mr. Ojeniyi asks for volunteers, and children loudly call out suggested alternatives. When called upon children jump out of their seats to rush to the blackboard.

One child announces, "ọ̀kẹ́ kan ó dín erinwó ó lé okaàn dínláààádọ́rin"; another announces "èèdẹ́gbàáwàà ó lé ẹgbẹ̀a ó lé ọ̀kàndínláààádọ́rin." A third youngster offers "ọ̀kẹ́ kan ó dín ọ̀tadínírinwó ó lé mẹ́sán." All these answers are correct, they all name the number "nineteen thousand six hundred and sixty-nine" but each indicates that the child got to that number by a different arithmetical path. Mr. Ojeniyi asks the class which of these is to be preferred. Few have any doubt that the first version is best, but Mr. Ojeniyi insists maybe not. Writing on the black board he demonstrates a notion of elegance in Yorùbá numbers using decimal number symbols to show the sequence of arithmetical operations mapped out in each named version, a form of modelling. Arithmetically these are three distinct versions of the same number.

- First: 19,669 translates as $(((20{,}000{\times}1){-}400){+}({-}1{-}10{+}(20{\times}4)))$
- Second: 19,669 translates as $((20{,}000{-}1{,}000){+}((200{\times}3){+}(20{\times}3)){+}9)$
- Third: 19,669 translates as $((20{,}000{\times}1){-}({-}(20{\times}3){+}400){+}9)$

He explains that he prefers the second because the first and the third use the name of a whole (ọ̀kọ́ kan, but then proceed to pull that whole apart. In his opinion the second version better shows how a number as a whole is its precisely defined parts, but he insists there is no absolute right or wrong version. A discussion ensues about how we can decide which form of the number might be best and why and what that tells us about number. The game Mr. Ojeniyi has the children playing is all about division of whole numbers—parts and wholes. I now understand his opening statement. I am awe-struck by this brilliant lesson in arithmetical calculation and translation between the Yorùbá number and English language metrics. I see too that it was the pure delight that Mr. Ojeniyi's feels for the aesthetics and patterns traced out by numbers, that is the inspiration for this class.

This ethnographic story which arose out of a slightly disconcerting surprise experienced by the author, offers a form of evidence. But in attesting that "this happened," the story is not evidence for a scientific truth claim: it is not proclaiming something about Yorùbá children, nor schools, nor contemporary Yorùbá life in general. The work the story is doing is more of a literary nature. It is setting out the particular situation in and from which this chapter is written. In that sense it is a framing device. But as part of an chapter in social study of numbers, the story is also doing some epistemic or knowledge work. It offers a beginning set of data, it is "a sighting" so to say, of Yorùbá numbers *in situ*, numbers in action along with the numbers of the modern standard decimal system, in an arithmetic lesson. It gives us—as reader and author—a point of departure.

3   Arithmetical Practices in Number Formation

In the Yorùbá number series there are fifteen basic numbers from which an infinite series is derived. I list these in Table 7.1 in a standard counting form.

The elaborated tone marking on the words I have listed above, point to the aural pleasure that can be derived from listening to Yorùbá number names being used, or for many speakers, even in their being read. The words make up a sort of music score. Spoken Yorùbá numbers are melodious like all Yorùbá words, and the tones-sounds offer clues in discerning the histories of the words. I explain the linguistic origins of these words naming the basic set of numbers in my next section, here I merely note that verbs—words denoting action—lie at the core of these names and in that they differ from the English set of basic numbers which are nouns and name things (for example "five" is a form of a very old noun meaning "hand"). And further, the sorts of action Yorùbá verbs denote is rather different than action as remarked by English verbs. Of rope

TABLE 7.1   The basic set of Yorùbá number names (Abraham 1962)

Set 1	ókan	èjì	ẹ̀ta		èrin	àrùún	ẹ̀fa	èje		èjọ	ẹ̀sọ̀ń	ẹ̀wàá
	one	two	three		four	five	six	seven	eight		nine	ten

Set 2	ogún	ọgbọ̀n
	twenty	thirty

Set 3	igba	irínwó	ẹ̀ẹ̀dẹ́gbàáwàá or ọ̀kọ́ kan (as whole)
	two hundred	four hundred	twenty thousand

one might say in Yorùbá, *ó gùn* (that which is characterised by being individ-uated longs; it longs; it is in a state of being long); of a container of palm-wine one might say *ó tobi* (that which is characterised by being individuated bigs; it bigs; it is in a state of being voluminous). A plot of land may be spoken of as *ó fẹ̀* (that which is characterised by being individuated spreads; it spreads; it is in the state of having a large surface), and of a nugget of gold, one might say *ó wúwo* (that which is characterised by being individuated heavys; it heavys; it is in a state of being heavy). In Yorùbá language usage, verbs 'read' action differ-ently than English, and by implication the sorts of things that are said to act in Yorùbá are different. But I am getting ahead of myself here.

Twenties matter in Yorùbá numbers. In counting, after twenty the core arith-metic process in the working of the Yorùbá metric is progression in jumps from whole vigesimal (twenty) point to the next whole twenty while noting the mul-tiplication factor. Then, when you get near to the number you are targeting, you utilise the secondary base of ten, and then in a third or tertiary stage, the further subsidiary base of five comes in. Instead of the addition of ones, addi-tions of tens, and addition of hundreds and so on, the process that is familiar in the decimal metric, in the Yorùbá metric it is multiplication which generates multiples of *ogún* (twenty). To be more exact, the Yorùbá verb embedded in a number in action in measuring means: "multiple placings." Working around the vigesimal points, intermediate numbers are then generated using tens and fives with strategic subtractions and additions.

Here is a general description of how to derive Yorùbá numbers. The vigesimal points occur at twenty, forty, sixty and so on, we can understand the generation of numbers up to sixty in the following way:

The first four numbers of each vigesimal are generated through a process which is fairly familiar to base ten users through the addition of ones, say:

$40 + 1 = 41$,

$40 + 2 = 42$, etc.

After 44, we progressively take away one less at each step to generate 45 to 49. We "leap" to:

$60 - 10 - 5 = 45$,

$60 - 10 - 4 = 46$ etc.

Then to generate 50 to 54:

$60 - 10 = 50$,

$60 - 10 + 1 = 51$ etc.

From 55 we progressively take away one less at each step to 59:

$60 - 5 = 55$,

$60 - 4 = 56$ etc.

Yorùbá numbers emerge as a nested cluster. Arriving at a particular number form by working from the nearest vigesimal can be imagined as setting out and aligning of parts, and of course, "passage" can be made differently through the relations of the ten and the five, different setting-out and aligning of parts is possible.

Yorùbá numbers were first written down as words, as number names, in the late nineteenth century; they became a series of inscribed names in the process of being collected as "a cultural thing" by European anthropologists, missionaries, linguists, and teachers. In the nineteenth century, Yorùbá numbers became objects of knowledge in anthropology. The numbers which had existed solely as uttered names, part and parcel of, indeed often the sociocultural pivot of, actual on-the-ground processes of counting and measuring by Yorùbá people, came to life in a different form—they became "modern" so to say. An integral part of that collection process was translation between two different metrics, and that process of writing down was part and parcel of understanding their working as a modern metric.

Devising numbers for counting and measuring in Yorùbá as part of social transactions, is arithmetically complicated, but it is also arithmetically efficient since with three bases, and strategic use of addition and subtraction, routes utilising different factorial patterns are possible. Up until a short time ago, only a few number names were set as standardised, and in this the Yorùbá metric was quite unlike the decimal system. Even a few years ago most of the larger numbers in Yorùbá were derived by experts on the spot, depending on the social demands of the particular situation. In the past, in any particular context the particular way of saying a number was settled upon as an expression of the sociocultural pushes and pulls that characterise any particular situation. Although this pressure to find exactly the right form of number for the right social situation is no longer so strong, there are still always worse and better ways to derive a Yorùbá number. There is an art to calculating in Yorùbá, although as computer mediated translation begins to become common in contemporary Yorùbá life, number forms are becoming much more standardised. Algorithms invariably deliver the same form of a number in writing or talking, so norms for number names are fast being established.

I end this section detailing the arithmetic involved in generating Yorùbá numbers, with the issue of well-intentioned attempts to "decimalise" the Yorùbá numeration system. A scheme of reform was developed in 1962 by Robert Armstrong when he was Director of the Ibadan Institute of Education, in which "all subtractive numerals are abolished, and a zero is added at the beginning ... [so that] any of the operations of arithmetic can be easily expressed verbally in this [proposed] decimal system" (Armstrong 1962, 21). In 1986, in commenting on my critical response to Armstrong's proposal (see Verran 2001, 250), anthropologist Karin Barber noted that:

> The education authorities have tried to "simplify" the Yorùbá numeral system by removing one stage in the process, i.e. the fives-up-and-down between the decades. Now they are copying the Indo-European system and adding from one to nine from the previous decade ... At the same time they've simplified the construction of the words ... As they've left the construction of decades intact I don't know whether the new system will make it easier or just confuse things further. ... The bridge between the two conceptual systems they were beginning to teach before I left is only a bridge ... at a very superficial level.

In considering to what extent reform of Yorùbá numeration is needed, and likely to foster both the survival of Yorùbá forms, and help contemporary Yorùbá children as they work in the modern world, we should remember that like any cultural product, number systems evolve. However, we also need to note that when it comes to numbers, differences between Yorùbá and English are large and pervasive. There is a need to ensure the continued coherence of the Yorùbá metric since it remains in wide usage, and when it comes to learning/teaching, there is a need to guard against the possibility of inadvertently introducing cognitive dissonance in inventing a "reformed" numeration system. These imply that "improvements" such as those proposed by Armstrong should be treated with great care.

In finishing this section, I stand back to get an overview of the patterns of relations traced out by the arithmetic characteristics of the two metrics. As part of that I speculate on the two metrics' sociomaterial origins, suggesting that the human body has been inspirational in the emergence of both. I see the normal digital complement of humans (fingers and toes arranged on hands and feet) is expressed in both, albeit construed quite differently. In Yorùbá it is the whole human being who figures with their twenty digits: four sets of five; two sets as hands, and two sets as feet. This four by five set, also seems to be a significant pattern in Yorùbá religious life (Segla 2017), and indeed four by five was also an important pattern in the cowrie shell currency in Yorùbá trading up until the end of the nineteenth century (Johnson [1921] 2010). I detect a circular moment in the Yorùbá metric, a circling inwards in engagement with the modalities of a complex sociomaterial present.

In English many have speculated that it is fingers that matter, that it was in acts of representing the world through fingers, say noting the passing of a single sheep through a gate by gesturing with a single finger raised or lowered, that numbers arose—ten suggesting that it is hands and eyes working together representing a cognised world. Numbers in the medium of fingers raised and lowered, or scratched markings on wood, as distinct from the messy complex world of sheep and their shepherds (Ifrah 1985). In this decimal patterning, singularity of fingers affords a gesture and the position in the series of gestures, little events, is what matters. This singularity and positionality contrasts with what is afforded by the four sets of five which comprise the whole human digital complement, the sense seemingly summoned up by the Yorùbá number pattern.

That sense of decimal numbers effecting a shift to an alternative medium of communication fits with the usual claim that the modern decimal one-to-many metric effects abstraction—a carrying off to another domain in translation to an alternative sociomaterial medium of communication. In such a shift what the number preserves is a relation, position in a metric scale of value. Such shifting in sociomateriality to an alternative communicative domain would in all likelihood be associated with different political and social stakes and different stakeholders. In contrast to that sense of abstraction, the Yorùbá whole-to-parts metric seems to embed a circling inwards, a staying with the here and now of the metric's use-in-place to order the here and now of a domain of transaction. While there is no sense of translation there is a strong sense of matter being (re)purposed with number and as number.

4   Linguistic Practices

In linguistic and philosophical studies of number there is controversy over the relation between language and number (Hurford 1987). Is number emergent within word usage, expressing the differing grammars arising in linguistic mechanisms? Or is it the other way around? Does number arising directly as cognitive mechanism drive the emergence of grammar? It is an ancient and much debated question and beyond the scope of this chapter to consider. It seems likely that both claims are right to some degree.

In skirting around this vexed issue let me instead announce that in my framing numbers are treated as words, and as such as expressions of the patterns of relations traced out in the grammars of particular languages. Thus, the Yorùbá metric, just like the Yorùbá language in its various dialects is an outcome of the forms of life that have emerged in particular places and times. Both metric and grammar express patterns of relationality, as indeed does the decimal metric and the grammar of English and its cognate languages. This does not mean that the metrics are incommensurable. Translation is feasible and indeed routine as we have already seen. As I noted in beginning the previous section, key in making good translations is recognising differences in the meaning making work of verbs in Yorùbá and English. In English one might say of a section of rope: "It is long", meaning something like "That pile you see coiled (here and now) is a long rope." One could add, "Actually it is ten metres long". In Yorùbá in the same situation pointing at the coils one might say "Ó gùn" (literally "It longs") and in providing detail, "*Okùn gùn mẹ́wàá mita* (literally: "Ropematter longs in mode divided, and in metre (*mita*) mode collected, ten; Or, in a better translation "The rope is ten metres long").

In following numbers in a sociomaterial inquiry, what is interesting is the ways this difference in the meaning making forms of verbs shows up in patterns of relations embedded and variously expressed in particular situations, mediated by numbers. Discerning and learning to "read" the ways the metrics sometimes mesh and sometimes interrupt each other, helps analysts develop a "feel" for the dynamic of a form of life. This for me is the sort of useful work that social study, sometimes called philosophical anthropology of numbers, can offer.

In English language usage the number "ten" in the sentence "I cooked ten potatoes" tells the listener about extent; literally the size of what will be available for supper. To say it more formally the statement refers to the extent of the quality of numerosity held by that group of potatoes sitting in water boiling in a pot in the hob. Numerosity may be a word you have never come across before, but this concept lies at the core of all modern decimal counting. It refers to the attribute that things have in being a thing, we might call it thingness; it is one of the qualities or properties held by a potato and by a collection of potatoes, which of course also have mass, or weight. I could have said "I cooked a kilogram of potatoes", and that would indicate a different quality held by potatoes. "Ten potatoes" names the extent to which numerosity is held; there are ten single brownish lumps; the potatoes in the pot on the stovetop hold the property of numerosity to the extent of ten. Each single potato exhibits that property to the extent of one. Similarly, in "She is one point five metres tall," the number "one point five" tells us about the extent of the quality or property of length in the body of a particular woman.

In English, numbers qualify in a second order way, they qualify a qualifier which names a property of an object. Of course, there are also numbers that tell about position in an order, the ordinal numbers, and other ways of using numbers, but here I am not aiming to give an exhaustive account of numbers and their usages, but rather to focus-up some obvious contrasts and connections between English and Yorùbá language metric usage that might be felt. I want to develop some insight into sociomaterial experience of numbers and their usage, and in this section I do that by probing the different ways numbers are used with words in talk and writing, in English and Yorùbá. This will involve me in using some grammatical concepts that readers might not be familiar with.

Readers are probably comfortable with the idea that sentences have subjects named by nouns, verbs naming actions, and objects that are somehow situated in the action. In both English and Yorùbá such object nouns are named by nouns that follow the verb, which we see above in "I cooked ten potatoes." A translation into Yorùbá would be "*Mo ṣe poteto mẹ́wàá*" (literally: "I made potatoes collected to the extent of ten"). We can parse these sentences: "I/Mo (subject, a pronoun) cooked/ṣe (verb) ten potatoes/*poteto mẹ́wàá*" (object, noun). I use this idea that sentences have distinct parts in what follows, but I will use more precise terms than subject, verb, and object, since while Yorùbá sentences usually have these similar parts, there are also profound grammatical differences between the languages that matter when it comes to discussing how numbers sit in language use and how that is sociomaterially significant. I will refer to sentences as having designants (roughly equivalent to subjects of sentences), and predicates (which include verbs and their objects).

In designating, in having subjects in our sentences, we talk of what comprises what we might call the real world; potatoes we are preparing to eat are one such. But in referring to potatoes in English we would assume they are first and foremost *things*, lumps of matter that have the qualities that characterise potatoes. "The potatoes are in the pot" we might say. Here the "the", in grammatical terms an article, provides a clue that English speakers designate spatiotemporal entities routinely in language—the subject of a sentence is imagined as a distinct and separate "thing." Those spatiotemporal entities routinely designated in English, indicated in this sentence by the "the," need to be qualified in measuring or quantifying. In using a number word in English a quality is first identified, and the extent to which that quality is held is remarked by the number word. With respect to the potatoes being prepared for supper, I previously indicated this as either ten, which points to the thingness (numerosity) of the spatiotemporal entities in the pot, or a one which indexes the mass as (kilogram) of the gathered together spatiotemporal entity in the pot.

An appropriate translation into Yorùbá of the English sentence "The potatoes are in the pot" would be *"Àwọn poteto wà ninú ìkòkò"*, but that sentence does not remark exactly what the English sentence does. A literal back-translation would be "That (very) potato matter (about to be cooked) exists inside the pot". There are no articles in Yorùbá grammar (no equivalent of "the" or "a" in English), so although "*àwọn*" seems to be "the," it is actually an emphatic pronoun that introduce; it implies "that" (a pointing-at, indicating that which is about to be cooked). It emphasises the noun "*poteto*" (a loan word from English) that is at issue here. This pronoun effecting emphasis is necessary because what is routinely designated in Yorùbá sentences (the subject), what the verb says will be acted upon, is *a sortal entity—"matter of potato-sort."* Matter of a particular sort is routinely conjured up in Yorùbá sentences, if one wants to say something about a particular sort of matter here and now in predicating something of it, then it is the manner in which that sort of matter manifests that will be remarked on. This is a very significant issue when it comes to the differences between Yorùbá and English number words in terms of how they are made. It explains why Yorùbá number words are elisions of introducers and verbs, why the names of the basic or "counting" set of Yorùbá numerals have the form of nominalised verb phrases.

Saying that the numerals are verb phrases that have been elided to form single words, identifies that the numerals function as mode or modal nouns in a grammatical sense. "Mode" here is used to point to the manner in which something manifests or is presented—how it is bundled. One could say in English "The product they were selling was presented in a unified mode, despite its several internal parts being obvious." The product being promoted in that situation could be taken in "a mode of one," or it could be (perhaps) "a mode of three distinct products." The crucial mode when it comes to numbers is dividedness: a number word in Yorùbá conjures up the extent of the dividedness of what is manifest. This is so because what is designated in a Yorùbá sentence (its subject), is a particular sort of matter which hypothetically exists vaguely as a whole distributed across spacetime. What a number word achieves in Yorùbá in naming a degree of dividedness manifest in a sortal entity is to indicate something about the form in which a particular sort of "stuff" actually happens to be manifesting in this situation, Yorùbá number words describe spatiotemporal form.

The grammar of Yorùbá language and the grammar of English have their speakers designating differently. And to say something about the subject of a Yorùbá sentence by using a number word, a speaker points to its form of spatiotemporal manifestation—the degree of its dividedness. But when a speaker of English, who has just designated a spatiotemporal entity as the subject of their sentence, wants to add some further information in the predicate, they point to the sort of stuff it is and note the extent of that sortality.

Let me now look a little more closely at how we see the modalising work that Yorùbá number words do in language use by teasing out some particular terms that are shortened and elided to form number words in use. This analytic technique uses the rules of elision to unpack, these are linguistic norms that are well established, and widely known to Yorùbá language users. In further elaborating the linguistic uniqueness of Yorùbá number I am looking inside the quantifying number words, to discern the linguistic resources used in their formulation. This exercise is made possible by a remarkable dictionary first published in 1946 by linguist Roy Clive Abraham (1890–1963) with a second edition in 1962 (Abraham 1962). In Yorùbá life there are four distinct sets of numbers each set denoting a different situation in which numbers are used. And each of these sets includes particular elided terms. Before I consider these different sets, each of which points to purposes, I return to look into the core number set I discussed briefly in the previous section. The core set is the arithmetical set, this is the form used in teaching. The number words are verb phrases and identifying the verbs involved helps us to recognise that Yorùbá number words name with precision particular sorts of relations achieved in certain actions. Through applying the rules for elision and vowel harmony in Yorùbá, it is possible to tease the elements of the number words apart, etymologically significant insights can be gained.

Apart from the primary set of names elaborated in Table 7.1 where origins cannot be discerned, the numbers of this core arithmetic set are formed from elisions involving three different verbs. When twenties are involved the verb which becomes part of the elided verb phrase, is ó nòn (it places out). So ọgọ́ọ̀ta, sixty (20x3, twenty placed out to three) arises this way: *ó nọ̀n* becomes *-ọ̀-*, which joining with *ogún* (twenty) elides to *ogún ọ̀ ẹ̀ta,* then by further elision becomes *ogún ọ̀ta* and in a final step *ọgọ́ọ̀ta*. The verb pointing to the 'placing out' recognised at the core of the word in the form of the *-ọ̀-*. *Ó dín* (it reduces) is the verb which signals subtraction. Fifty for example is signalled as ten removed from sixty (–10+(20 x 3)), seventy as (–10 + (20 x 4)–10), ninety as (–10 + (20 x 5)). Each of these number words is prefixed by *àádín* elision of *ẹ̀wàá ó dín* (ten it reduces), thus with further elision fifty is *àádọ́ọta* (*àádín ọgdọ́ọ̀ta* by way of *àádọ́gọ́ọ̀ta*). *Ó lé* (it adds to) is a third verb phrase involved. So, eleven is *ókọ̀nlàá* (1 + 10) elision of one (*ókọ̀n-*) adds to (*-lé-*) ten (*-ẹ̀wàá*).

Passing on to consider further forms of Yorùbá number names, let me briefly point to two further verb forms deeply involved with number usage in Yorùbá, when the core arithmetic set is used to count something that is indicated. If it is currency, that is indicated by adding an elided form of the noun for money, *owó* into the number word. When counting other things, a form of *mú* an obsolete verb related in meaning to the present day *mún* (to take or pick up several things in a group or as one) is used. The core number word form (already a mode noun—an elided phrase with verb and introducer) is further modified when used in quantification statements in Yorùbá talk. In the collection of things set, the core number name is prefixed with "m" and a high tone. Thus, the derived *ajá méjì* (two dogs): the number word here implies literally "dog-matter in the mode of being grouped in the mode of being two"; or I might say say "*Ó fún mi ni ìwé mẹ́rin*" which is conventionally translated as "He gave me four books". A more literal translation is "He gave me bookmatter in the mode of a group in the mode of four".

As a further set there is an ordinal set of number words focussing up positioning and ranking, which tell of the number of times the event of a particular manifestation has occurred. There is a noun involved here *ìgbà* (a time) as in *nûgbà yìí* (at the present time), when attached to a number word this become elided to *ẹ̀ẹ̀* and the quasi verb *kọ́* is included. Coming together this generates a highly elaborated form which is a modality of a modality of a modality. A literal translation of "*Ó gbà ìwé ìkẹta ni* (He took the third book)" illustrates the triple modal nature of this form of numeral: "He took bookmatter in the mode of collected together individual items, in the mode of three, as the third positioned" mode noun formed by nominalising a verb phrase that already contains a mode noun.

This section of the chapter has offered analysis of how the different numbers we met in the previous section, work as words in Yorùbá and English. Likewise, considering relations between numbers and words is the final section of my book *Science and an African Logic* (2001). Indeed this paper can be considered a short version of that book. Beginning with an analytic ethnographic story about numbers in use in an arithmetic lesson in Yorùbáland, which I noted subsequently had me talking to a large number of children, asking them to tell me how they did that, I followed up with a different sort of analysis, plotting out the arithmetical processes that differentially make Yorùbá numbers and English numbers. Then this section has been a third analysis laying out how numbers work differentially in meaning making using Yorùbá and compared to English. In this section I showed how the alternatively focussed meaning making work of Yorùbá and English verbs matters when it comes to numbers.

Associated with the different meanings that verbs, as words in some sense re-performing action, carry, Yorùbá has its speakers designating different types of entities in talk and writing than English. Yorùbá routinely designates matter of a particular sort, the subjects of Yorùbá sentences are characterised by have par-ticular sorts of *qualities*, what is presented in Yorùbá language is a world where things have particular characteristics. In contrast English routinely designates the spatiotemporal situation of matter, the world's material *form* is focused up as what exactly it is that is being talked about, that is, being designated. In this sec-tion I have shown, that perhaps unexpectedly, this difference in verb meaning making matters when it comes to how numbers sit in and work in language use.

5   Making Something of the Metrics' Arithmetic and Linguistic Differences

In beginning to answer the "So what?" question that a puzzled reader who has by now waded through a great deal of perhaps tedious detail might justifiably ask, I go back to Mr. Ojeniyi in his classroom faced with a group of attentive children. He is an experienced teacher and is aware that the children sitting attentively before him likely have a lot of practical experience with numbering. In beginning to show how interpretation might make use of the detailed arith-metic and linguistic description of the internal workings of the dual metrics of contemporary Yorùbá life I have just given, I return to my ethnographic story. I identified the story as a framing and as offering a beginning in interpretation of the experienced actuality of contemporary Yorùbá life involving numbers, an exegesis, or perhaps more precisely, an eisegesis, a reading into the experience of the lesson had by the ethnographer as revealed in the story.

I begin with the figure of Mr. Ojeniyi and in particular his expressed prefer-ence for one of the Yorùbá versions of the decimal number nineteen thousand, six hundred and sixty-nine. He announced that he preferred "*ẹ̀ẹ́dẹ́gbàáwàá ó lé ẹgbẹ̀a ó lé ọ̀kàndínláàádọ́rin*", rejecting those numbers whose names begin with *ọ̀ké kan*, twenty thousand for numbers that have a lesser value than twenty thousand. His reasoning was that as a precise whole number in Yorùbá, *ọ̀ké kan* is a whole value in Yorùbá and that wholeness is significant in the workings of the metric. Still he accepted this preference as based on a feeling that the whole of twenty thousand should be respected as such, as a recognised whole.

Putting myself in the shoes of Mr. Ojeniyi, as I wrote and honed the ethnographic story I came to see him as meaning something like this. If you use the rationality of the decimal metric all three translations the children offered are correct. If you use the rationality of the Yorùbá metric the alternatives which start with *ọ̀ké kan* are not acceptable. He was clearly recognising both rationalities as equivalent, while urging the children to apply the conventional standards of each of the metrics appropriately: know the alternative standards and respect them by developing habits of working Yorùbá numbers that respect them. I take away two insights from listening carefully to Mr. Ojeniyi, from paying close attention as a learner in his lesson.

First, Mr. Ojeniyi insists that there are rationally alternative ways of knowing numbers, and goes to some pains to have the children in his class learn these rationalities well. Second, he insists that one can feel, can experience numbers, and he urges the pupils to develop such feelings and take notice of them when they are using numbers. While I never did develop a technical capacity for fluently using the rationality of Yorùbá numbers, as a learner I did take to heart both the existence of the dual rationalities as expressed in metrics, and with that awareness gradually honed my capacity to feel for different sociomaterial expressions of numbers.

Mr. Ojeniyi is certainly amongst good company in asserting these as realities of conceptual life. The eminent American cognitive scientist Susan Carey (2009) is equally explicit on these two issues. She differentiates the rationality of the metric that is the set of numbers up to somewhere between five and ten as expressing a rationality embedded in biology and evolution. The distinction here can be seen in a comparison between five and say fifty-five, which she points to as having a cultural origin. While both five and fifty-five are said, in English, to be cardinal numbers indexing value in the physical stuff of the world, indexical numbers—the rational basis of their being—so differs. Carey argues that such differentially felt contrasts arise in concepts' alternative origins.

There is another distinction between numbers that can be felt: that between cardinality where numbers work as indexes in meaning making; and ordinality where numbers work as icons in meaning making. Indexical practices begin in valuing with social orderings effected through relations between values. The alternative sequence holds in numbers being iconic, they begin in ordering and end by valuing. The difference felt concerns what we might call the stickiness of the felt connection between the social and the materiality. Indexical numbers and iconic numbers express alternative modes of the sociomateriality, felt as alternative rationalities. With indexical numbers socialities might easily be separated off from materialities, but not so in iconic numbers where sociality and materiality become strongly entangled in the practices of their constitution.

Having taken Mr. Ojeniyi's lesson seriously, learning from it and from others delivered by his colleagues, as I elaborate in *Science and an African Logic*, I sought clarification from children themselves. I arranged to talk rather intensely to a large number of both English and Yorùbá speaking children, around half of who were profoundly bilingual. I discovered that the bilingual children already knew what I had just discovered. I felt myself extremely fortunate that many of these children, sensing my deficit in such matters, went out of their way to explain to me how they did it. Here is an account of what some of them said.

Imagine Afolabi watching as a small tin, which had formerly contained condensed milk, is filled with peanuts and emptied into a wide shallow plastic bowl, and then filled a second time and emptied into a thin tall glass mug. I ask him if the peanuts in the bowl and the cup are the same amount. Speaking English, he says, "The peanuts in the bowl and the mug, they look different. But that are the same: one tin here and one tin there." Afolabi attributes the quality of thingness (a tin of nuts) to the different presentations of peanuts in the wide flat bowl and the thin tall mug. His judgement is correct, whereas the younger child I had just spoken to was convinced that the tall mug had more nuts. Toyin, who is eleven knows too that you can solve the puzzle if you think of any "collection of stuff" as a thing. Nevertheless, she tells me in English you can think about it differently too: "When I look at it one way they look the same, then when I look at it another way they don't look the same. That's when I think 'It's just a tin of nuts', but when I think about the bowl I can see that if I push way [indicating with her hands the diameter of the bowl], it will make them higher; it's the same as in the cup." Toyin is attributing two different qualities to the two collections of peanuts: thingness ("a tin of nuts") and volume, noting that the amounts in the bowl and mug are equal. She knows too that to think about volume you need to visualise its dimensions. Toyin could even comment on the usefulness of the different qualities. I was asking Toyin if she would think that the peanuts in the mug were equal in amount to the peanuts in the bowl if she had not watched when I emptied the tin. She replied that "If you're thinking about the space that the peanuts fill up, you can try to imagine if they will look the same when the peanuts in the bowl are squashed up the same way as they are in the cup." In other words, Toyin is focusing on what we might call the the *spacefillingness* of the peanuts. She explains that if you look knowingly at the material presented to you, then you can estimate amounts.

'Bola is a village child, while she learns English at school, she is function-ally monolingual in Yorùbá. Here is what 'Bola (eleven years old) speaking in Yorùbá had to say. She watches as full tins of peanuts are emptied into the plastic bowl and the mug. I ask in Yorùbá if there are the same amounts of peanuts in the bowl and the cup: "*Ǹ jé iye hóró èpà kan náà ló wà nínú kóòbù yìí àti abó yìí?*" ("Is there the same amount of peanuts in this cup and bowl?") 'Bola laughs and replies, "*Òkan wà níbí òkan wà lóhùn-ún*" ("There is one here and one there"). I ask 'Bola if she is quite sure that there is the same amount of peanuts in the two containers, she almost scoffs, "*Eyo kan ni eyo kan, àfi tí a bá pin in si méjì bèè ni mo ni wò òokó pin in.*" ("One is one unless you divide it into two, and I watched and you didn't divide it"). Then I ask her whether she would know they were the same if she had turned her back while I poured the peanuts out: "*Tó bá se pé o wo eèhìn ni gbà ti mo ni da èpà náà ni, ò bá mò pé iye kan náà ló wà ninú kóòbù áti koto náà?*" She replies, "*Rárá ó seése ki o ti pin in ki o si ti mú díè lo tà fún elòmiràn*", ("No you might have divided it and taken some away to sell to another person".) When I ask 'Bola if the peanuts in the bowl and the cup look the same amount, she replies "*Won kò dógba*" ("They are not equal"). When I repeat the question putting emphasis on *iye* (amount), "*Ǹ jé ó dà bi eni pé iye kan náà ni wón,*" 'Bola asserts that you cannot know whether it is the same amount by just looking, "*Ó nira láti mò bóyá iye kan náà ni wón nipa wíwò*" ("It is difficult to know if they are the same size by looking").

Folake is bilingual and speaks and quantifies fluently in both Yorùbá and English. Here is Folake, aged nine, explaining in Yorùbá why the coke in a bottle is the same as that contained in a wide-mouthed plastic mug filled with the contents of a second bottle of coke. "*Ara kan náà ní wón tórí pé inú ìgò kékéré náà ni won fi si, o si jé kí o jó èyìí sùgbón àpapò èyìí àti èyìí jé òkan náà.*" ("They are the same because they put this there in this little bottle and that made them look like this. But the aggregate of this one [indicating the differ-ence in width of the two containers] and this one [indicating the difference in the two heights of the liquid] is the same one.") Although she does not name the quality involved in English language measurement of liquids (volume), in connecting up the unitary one involved Yorùbá measurement with a sense of the volume of the liquid, Folake is prepared to comment on the nature of her unitary feature, indicating that it is a unit of *spacefillingness*, but she still talks of it as a one. Folake is one of the bilingual children who is confidently connecting the cognitive processes of measuring in Yorùbá with measuring in English.

6   Concluding

In working my way to ending my chapter, and in keeping with the autoethnographic thread by which I have woven this chapter together, I shift the scene in which I imagine numbers as coming to life; no longer a Nigerian classroom but rather an Australian river whose health as a river is compromised, and a cause of worry amongst some citizen scientists in Melbourne. Having returned to my home country, in following what I saw as a significant shift in Australian environmental governance, I began to take a close interest in the numbers involved in that work.

The two insights into the lives of numbers that I had developed as a consequence of paying close attention in Mr. Ojeniyi's lesson, proved crucial in this new study. I felt grateful too to the kindly bilingual children who had helped me hone new understandings of numbers' lives. Thanks to those experiences, I had developed a capacity to experience different rationalities expressed in numbers, and these skills developed in making sense of living through dual metrics in Nigeria proved invaluable. They enabled me to inquire into the differences in number use when environmental water is cared for with science—indexical numbers, or traded in a newly instituted water market (Verran 2010). Having developed a capacity to feel the alternative rationalities of indexical metrics and iconic metrics in Nigeria was crucial this new analytic work.

Now in concluding this chapter, I use that insight developed in studying numbers as they are used in managing Australia's environmental water through water trading. I use this to throw new light on the insights I developed under the tutelage of Mr. Ojeniyi and the Yorùbá children. I go one step further in interpretation than in previous commentaries on the dual metrics of Yorùbá life. I propose Yorùbá numbers can be understood as primarily iconic numbers. Of course however, by adding an extra step in the measuring process, they can be rendered in ways that have them working as indexical numbers. This contrasts with decimal numbers as taught in primary school classrooms. There iconicity is treated as a minor rationality, more or less irrelevant to the sociomaterial work of numbers. But as we have seen decimal numbers working as icons is crucial in financialisation.

I am suggesting that Yorùbá numbers as icons, intrinsically carry the feel of numbers agential in trading in a way that those decimal numbers taught in primary school classrooms as indexes do not. In having ready access to the differing rationalities of both the metrics that are significant in contemporary life, and in being trained in classrooms in using both metrics, I propose Yorùbá children become equipped equally as entrepreneurs, where iconic numbers matter, and as experimental scientists where indexical numbers are crucial.

Bibliography

Abraham, R.C. 1962. *Dictionary of Modern Yorùbá*. London: Hodder and Staughton.

Akinadé, Olúgbénga O and Ọdẹ́túnjí A. Ọdẹ́jọbí. 2014. "Computational Modelling of Yorùbá Numerals in a Number-to-Text Conversion System." *Journal of Language Modelling* 2, no. 1: 167–211.

Armstrong, Robert, G. 1962. "Yorùbá Numerals." *Nigerian Social and Economic Studies*, no. 1. Ibadan: Oxford University Press.

Barber, Karin. 1984. *Yoruba Dun Un So: A Beginner's Course in Yoruba*. New Haven: Yale University Press.

Carey, Susan. 2009. *Origin of Concepts*. Oxford: Oxford University Press.

Hurford, James, R. 1987. *Language and Number: The Emergence of a Cognitive System*. Oxford: Basil Blackwell.

Ifrah, Georges. 1985. *From One to Zero: A Universal History of Numbers*. New York: Viking Penguin.

Johnson, Samuel. (1921) 2010. *The History of the Yorùbás: From the Earliest Times to the Beginning of the British Protectorate*. Cambridge: Cambridge University Press.

Segla, Dafon. Aimé. 2017. "Culture and Cosmos: Structural Changes In A System Of Knowledge-History, Concepts And Logic In Yorùbá Number Concept Development From People's Conception Of The Sky (West Africa)." *Revue des Sciences du Langage et de la Communication* 4, 360–75.

Verran, Helen. 2001. *Science and an African Logic*. Chicago: Chicago University Press.

Verran, Helen. 2010. "Number as an Inventive Frontier in Knowing and Working Australia's Water Resources." *Anthropological Theory* 10, no. 1: 171–78.

Verran, Helen. 2015. "Enumerated Entities in Public Policy and Governance." In *Mathematics, Substance and Surmise*, edited by Ernest Davis and Philip Davis, 365–79. Switzerland: Springer International Publishing.

Index

Printed in the United States
by Baker & Taylor Publisher Services